Δ-Formulas Dealing with Deductions

D0212821

Formula	Defined on Page:	Interpretation
$AxiomOfN(a)$	167	a is the Gödel number of an axiom of N.
$LogicalAxiom(a)$	168	a is the Gödel number of a logical axiom.
$QRRule(c, e, i)$	171	c codes a sequence of formulas (which is supposed to be a deduction), e codes the ith formula of the sequence, and e is justified by the quantifier rules.
$PrimeComponent(u, f)$	172	u and f are Gödel numbers of formulas, and the formula coded by u is a prime component of the formula coded by f.
$PrimeList(c, i, r)$	173	r codes the prime components in the first i entries of the deduction coded by c.
$TruthAssignment(c, i, r, v)$	173	v codes an assignment of truth values to the prime components coded by r.
$Evaluate(e, r, v, y)$	175	e is the Gödel number of a formula α, and y is the truth of α, given v, a truth assignment for r, which is a prime list that includes the prime components of α.
$PCRule(c, e, i)$	175	e is the ith entry of the deduction coded by c, and is justified by the propositional consequence rule.
$Deduction(c, f)$	177	f is the Gödel number of a formula and c is a code for a deduction of that formula.
$Thm_N(f)$	179	f is the Gödel number of a theorem of N. **⟹Not a Δ-formula⟸**

42667909

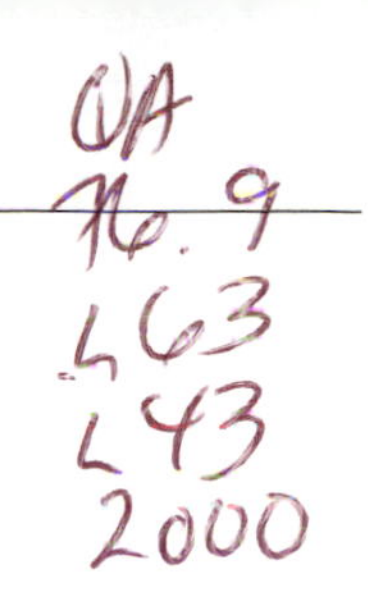

A Friendly Introduction to Mathematical Logic

Christopher C. Leary

State University of New York
College at Geneseo

DISCARDED

NORMANDALE COMMUNITY COLLEGE
LIBRARY
9700 FRANCE AVENUE SOUTH
BLOOMINGTON MN 55431-4399

Prentice Hall, Upper Saddle River, New Jersey 07458

JUN 2 8 2001

Library of Congress Cataloging-in-Publication Data
LEARY, CHRISTOPHER, C.

A friendly introduction to mathematical logic/Christopher C. Leary.
p. cm.
Includes bibliographical references and index
ISBN 0-13-010705-0

1. Computer logic. 2. Logic, symbolic and mathematical. I. Title

QA76.9.L63 L43 2000 99-052220
005.1'01'5113--dc21 CIP

Acquisitions Editor: *George Lobell*
Production Editors: *Lynn M. Savino/Bayani Mendoza de Leon*
Assistant Vice President of Production and Manufacturing: *David W. Riccardi*
Senior Managing Editor: *Linda Mihatov Behrens*
Executive Managing Editor: *Kathleen Schiaparelli*
Manufacturing Buyer: *Alan Fischer*
Manufacturing Manager: *Trudy Pisciotti*
Marketing Manager: *Melody Marcus*
Marketing Assistant: *Vince Jansen*
Director of Marketing: *John Tweeddale*
Art Director: *Jayne Conte*
Cover Designer: *Bruce Kenselaar*
Editorial Assistant: *Gale Epps*

©2000 by Prentice-Hall, Inc.
Upper Saddle River, New Jersey 07458

All rights reserved. No part of this book may be reproduced, in any form or by any means, without permission in writing from the publisher.

Printed in the United States of America
10 9 8 7 6 5 4 3 2

ISBN 0-13-010705-0

Prentice-Hall International (UK) Limited, *London*
Prentice-Hall of Australia Pty. Limited, *Sydney*
Prentice-Hall Canada, Inc., *Toronto*
Prentice-Hall Hispanoamericana, S.A., *Mexico*
Prentice-Hall of India Private Limited, *New Delhi*
Prentice-Hall of Japan, Inc. *Tokyo*
Pearson Education Asia Pte. Ltd.
Editora Prentice-Hall do Brasil, Ltda., *Rio de Janeiro*

To Andreas R. Blass and to the memory of Edward T. Wong

Two wonderful teachers
Two wonderful friends

And to my family:

Sharon, Heather, and Eric Leary

Contents

Preface

This book covers the central topics of first-order mathematical logic in a way that can reasonably be completed in a single semester. From the core ideas of languages, structures, and deductions we move on to prove the Soundness and Completeness Theorems, the Compactness Theorem, and Gödel's First and Second Incompleteness Theorems. There is an introduction to some topics in model theory along the way, but I have tried to keep the text tightly focused.

One choice that I have made in my presentation has been to start right in on the predicate logic, without discussing propositional logic first. I present the material in this way as I believe that it frees up time later in the course to be spent on more abstract and difficult topics. It has been my experience in teaching from preliminary versions of this book that students have responded well to this choice. Students have seen truth tables before, and what is lost in not seeing a discussion of the completeness of the propositional logic is more than compensated for in the extra time for Gödel's Theorem.

I believe that most of the topics I cover really deserve to be in a first course in mathematical logic. Some will question my inclusion of the Löwenheim–Skolem Theorems, and I freely admit that they are included mostly because I think they are so neat. If time presses you, that section might be omitted. You may also want to soft-pedal some of the more technical results in Chapter 4.

The list of topics that I have slighted or omitted from the book is depressingly large. I do not say enough about recursion theory or model theory. I say nothing about linear logic or modal logic or second-order logic. All of these topics are interesting and important, but I believe that they are best left to other courses. One semester is, I believe, enough time to cover the material outlined in this book relatively thoroughly and at a reasonable pace for the student.

Thanks for choosing my book. I would love to hear how it works for you.

To the Student

Welcome! I am really thrilled that you are interested in mathematical logic and that we will be looking at it together! I hope that my book will serve you well and will help to introduce you to an area of mathematics that I have found fascinating and rewarding.

Mathematical logic is absolutely central to mathematics, philosophy, and advanced computer science. The concepts that we discuss in this book—models and structures, completeness and incompleteness—are used by mathematicians in every branch of the subject. Furthermore, logic provides a link between mathematics and philosophy, and between mathematics and theoretical computer science. It is a subject with increasing applications and of great intrinsic interest.

One of the tasks that I set for myself as I wrote this book was to be mindful of the audience, so let me tell you the audience that I am trying to reach with this book: third- or fourth-year undergraduate students, most likely mathematics students. The student I have in mind may not have taken very many upper-division mathematics courses. He or she may have had a course in linear algebra, or perhaps a course in discrete mathematics. Neither of these courses is a prerequisite for understanding the material in this book, but some familiarity with proving things will be required.

In fact, you don't need to know very much mathematics at all to follow this text. So if you are a philosopher or a computer scientist, you should not find any of the core arguments beyond your grasp. You do, however, have to work abstractly on occasion. But that is hard for all of us. My suggestion is that when you are lost in a sea of abstraction, write down three examples and see if they can tell you what is going on.

At several points in the text there are asides that are indented and start with the word *Chaff*. I hope you will find these comments helpful. They are designed to restate difficult points or emphasize important things that may get lost along the way. Sometimes they are there just to break up the exposition. But these asides really are chaff, in the sense that if they were blown away in the wind, the mathematics that is left would be correct and secure. But do look at them—they are supposed to make your life easier.

Just like every other math text, there are exercises and problems for you to work out. Please try to at least think about the problems. Mathematics is a contact sport, and until you are writing things down and trying to use and apply the material you have been studying, you don't really know the subject. I have tried to include problems of different levels of difficulty, so some will be almost trivial and others will give you a chance to show off.

This is an elementary textbook, but elementary does not mean easy. It was not easy when we learned to add, or read, or write. You will find the going tough at times as we work our way through some very difficult and technical results. But the major theorems of the course—Gödel's Completeness Theorem, the incompleteness results of Gödel and Rosser, the Compactness Theorem, the Löwenheim–Skolem Theorem—provide wonderful insights into the nature of our subject. What makes the study of mathematical logic worthwhile is that it exposes the core of our field. We see the strength and power of mathematics, as well as its limitations. The struggle is well worth it. Enjoy the ride and see the sights.

Thanks

Writing a book like this is a daunting process, and this particular book would never have been produced without the help of many people. Among my many teachers and colleagues I would like to express my heartfelt thanks to Andreas Blass and Claude Laflamme for their careful readings of early versions of the book, for the many helpful suggestions they made, and for the many errors they caught.

I am also indebted to Paul Bankston of Marquette University, William G. Farris of the University of Arizona at Tucson, and Jiping Liu of the University of Lethbridge for their efforts in reviewing the text. Their thoughtful comments and suggestions have made me look smarter and made my book much better.

The Department of Mathematics at SUNY Geneseo has been very supportive of my efforts, and I would also like to thank the many students at Oberlin and at Geneseo who have listened to me lecture about logic, who have challenged me and rewarded me as I have tried to bring this field alive for them. The chance to work with undergraduates was what brought me into this field, and they have never (well, hardly ever) disappointed me.

Much of the writing of this book took place when I was on sabbatical during the fall semester of 1998. The Department of Mathematics and Statistics at the University of Calgary graciously hosted me during that time so I could concentrate on my writing.

I would also like to thank Michael and Jim Henle. On September 10, 1975, Michael told a story in Math 13 about a barber who shaves every man in his town that doesn't shave himself, and that story planted the seed of my interest in logic. Twenty-two years later, when I was speaking with Jim about my interest in possibly writing a textbook, he told me that he thought that I should approach my writing as a creative activity, and if the book was in me, it would come out well. His comment helped give me the confidence to dive into this project.

The typesetting of this book depended upon the existence of Leslie Lamport's LaTeX. I thank everyone who has worked on this typesetting system over the years, and I owe a special debt to David M. Jones for his Index package, and to Piet von Oostrum for Fancyheadings.

Many people at Prentice Hall have worked very hard to make this book a reality. In particular, George Lobell, Gale Epps, and Lynn Savino have been very helpful and caring. You would not be holding this book without their efforts.

But most of all, I would like to thank my wife, Sharon, and my children, Heather and Eric. Writing this book has been like raising another child. But the real family and the real children mean so much more.

Chris Leary
Geneseo, New York leary@geneseo.edu

A Friendly Introduction to Mathematical Logic

Chapter 1

Structures and Languages

Let us set the stage. In the middle of the nineteenth century, questions concerning the foundations of mathematics began to appear. Motivated by developments in geometry and in calculus, and pushed forward by results in set theory, mathematicians and logicians tried to create a system of axioms for mathematics, in particular, arithmetic. As systems were proposed, notably by the German mathematician Gottlob Frege, errors and paradoxes were discovered. So other systems were advanced.

At the International Congress of Mathematicians, a meeting held in Paris in 1900, David Hilbert proposed a list of 23 problems that the mathematical community should attempt to solve in the upcoming century. In stating the second of his problems, Hilbert said:

> *But above all I wish to designate the following as the most important among the numerous questions which can be asked with regard to the axioms [of arithmetic]:* To prove that they are not contradictory, that is, that a finite number of logical steps based upon them can never lead to contradictory results. *(Quoted in [Feferman 98])*

In other words, Hilbert challenged mathematicians to come up with a set of axioms for arithmetic that were guaranteed to be consistent, guaranteed to be paradox-free.

In the first two decades of the twentieth century three major schools of mathematical philosophy developed. The Platonists held that mathematical objects had an existence independent of human thought, and thus the job of mathematicians was to discover the truths about these mathematical objects. Intuitionists, led by the Dutch mathematician L. E. J. Brouwer, held that mathematics should be restricted to concrete operations performed on finite structures. Since vast areas of modern mathematics depended on using infinitary methods, Brouwer's position implied that most of the mathematics of the previous 3000 years should be discarded until the results could be reproved using finitistic arguments. Hilbert was appalled at this suggestion and he became the leading exponent of the Formalist school, which held that mathematics was nothing more than the manipulation of meaningless symbols according to certain rules and that the consistency of such a system was nothing more than saying that the rules prohibited certain combinations of the symbols from occurring.

Hilbert developed a plan to refute the Intuitionist position that most of mathematics was suspect. He proposed to prove, using finite methods that the Intuitionists would accept, that all of classical mathematics was consistent. By using finite methods in his consistency proof, Hilbert was sure that his proof would be accepted by Brouwer and his followers, and then the mathematical community would be able to return to what Hilbert considered the more important work of advancing mathematical knowledge. In the 1920s many mathematicians became actively involved in Hilbert's project, and there were several partial results that seemed to indicate that Hilbert's plan could be accomplished. Then came the shock.

On Sunday, September 7, 1930, at the Conference on Epistemology of the Exact Sciences held in Königsberg, Germany, a 24-year-old Austrian mathematician named Kurt Gödel announced that he could show that there is a sentence such that the sentence is true but not provable in a formal system of classical mathematics. In 1931 Gödel published the proof of this claim along with the proof of his Second Incompleteness Theorem, which said that no consistent formal system of mathematics could prove its own consistency. Thus Hilbert's program was impossible, and there would be no finitistic proof that the axioms of arithmetic were consistent.

Mathematics, which had reigned for centuries as the embodiment of certainty, had lost that role. Thus we find ourselves in a situation where we cannot prove that mathematics is consistent. Although I believe in my heart that mathematics is consistent, I know in my brain that I will not be able to prove that fact, unless I am wrong. For if I am wrong, mathematics is inconsistent. And (as we will see) if mathematics is inconsistent, then it can prove anything, including the statement which says that mathematics is consistent.

So do we throw our hands in the air and give up the study of mathematics? Of course not! Mathematics is still useful, it is still beautiful, and it is still interesting. It is an intellectual challenge. It compels us to think about great ideas and difficult problems. It is a wonderful field of study, with rewards for us all. What we have learned from the developments of the nineteenth and twentieth centuries is that we must temper our hubris. Although we can still agree with Gauss, who said that, "Mathematics is the Queen of the Sciences . . . " she no longer can claim to be a product of an immaculate conception.

Our study of mathematical logic will take us to a point where we can understand the statement and the proof of Gödel's Incompleteness Theorems. On our way there, we will study formal languages, mathematical structures, and a certain deductive system. The type of thinking, the type of mathematics that we will do, may be unfamiliar to you, and it will probably be tough going at times. But the theorems that we will prove are among the most revolutionary mathematical results of the twentieth century. So your efforts will be well rewarded. Work hard. Have fun.

1.1 Naïvely

Let us begin by talking informally about mathematical structures and mathematical languages. There is no doubt that you have worked with mathematical models in several previous mathematics courses, although in all likelihood it was not pointed out to you at the time. For example, if you have taken a course in linear algebra, you have some experience working with $\mathbb{R}^2$, $\mathbb{R}^3$, and $\mathbb{R}^n$ as examples of vector spaces. In high school geometry you learned that the plane is a "model" of Euclid's axioms for geometry. Perhaps you have taken a class in abstract algebra, where you saw several examples of groups: The integers under addition, permutation groups, and the group of invertible $n \times n$ matrices with the operation of matrix multiplication are all examples of groups—they are "models" of the group axioms. All of these are mathematical models, or structures. Different structures are used for different purposes.

Suppose we think about a particular mathematical structure, for example $\mathbb{R}^3$, the collection of ordered triples of real numbers. If we try to do plane Euclidean geometry in $\mathbb{R}^3$, we fail miserably, as (for example) the parallel postulate is false in this structure. On the other hand, if we want to do linear algebra in $\mathbb{R}^3$, all is well and good, as we can think of the points of $\mathbb{R}^3$ as vectors and let the scalars be real numbers. Then the axioms for a real vector space are all true when interpreted in $\mathbb{R}^3$. We will say that $\mathbb{R}^3$ is a model of the axioms for a vector space, whereas it is not a model for Euclid's axioms for geometry.

As you have no doubt noticed, our discussion has introduced two separate types of things to worry about. First, there are the mathematical models, which you can think of as the mathematical worlds, or constructs. Examples of these include $\mathbb{R}^3$, the collection of polynomials of degree 17, the set of 3×2 matrices, and the real line. We have also been talking about the axioms of geometry and vector spaces, and these are something different. Let us discuss those axioms for a moment.

Just for the purposes of illustration, let us look at some of the axioms which state that V is a real vector space. They are listed here both informally and in a more formal language:

Vector addition is commutative: $(\forall u \in V)(\forall v \in V)u + v = v + u$.

There is a zero vector: $(\exists 0 \in V)(\forall v \in V)v + 0 = v$.

One times anything is itself: $(\forall v \in V) 1v = v$.

Don't worry if the formal language is not familiar to you at this point; it suffices to notice that there *is* a formal language. But do let me point out a few things that you probably accepted without question. The addition sign that is in the first two axioms is not the same plus sign that you were using when you learned to add in first grade. Or rather, it *is* the same sign, but you *interpret* that sign differently. If the vector space under consideration is $\mathbb{R}^3$, you know that as far as the first two axioms up there are concerned, addition is vector addition. Similarly, the 0 in the second axiom is not the real number 0; rather, it is the zero vector. Also, the multiplication in the third axiom that is indicated by the juxtaposition of the 1 and the v is the scalar multiplication of the vector space, not the multiplication of third grade.

So it seems that we have to be able to look at some symbols in a particular formal language and then take those symbols and relate them in some way to a mathematical structure. Different interpretations of the symbols will lead to different conclusions as regards the truth of the formal statement. For example, if we take the commutivity axiom above and work with the space V being $\mathbb{R}^3$ but interpret the sign $+$ as standing for cross product instead of vector addition, we see that the axiom is no longer true, as cross product is not commutative.

These, then, are our next objectives: to introduce formal languages, to give an official definition of a mathematical structure, and to discuss truth in those structures. Beauty will come later.

1.2 Languages

We will be constructing a very restricted formal language, and our goal in constructing that language will be to be able to form certain statements about certain kinds of mathematical structures. For our work, it will be necessary to be able to talk about constants, functions, and relations, and so we will need symbols to represent them.

Chaff: Let me emphasize this once more. Right now we are discussing the *syntax* of our language, the marks on the paper. We are not going to worry about the semantics, or meaning, of those marks until later—at least not formally. But it is silly to pretend that the intended meanings do not drive our choice of symbols and the way in which we use them. If we want to discuss left-hemi-semi-demi-rings, our formal language should include the function and relation symbols that mathematicians in this lucrative and exciting field customarily use, not the symbols involved in chess, bridge, or right-hemi-semi-para-fields. It is not our goal to confuse anyone more than is necessary. So you should probably go through the exercise right now of taking a guess at a reasonable language to use if our intended field of discussion was, say, the theory of the natural numbers. See Exercise 1.

Definition 1.2.1. A **first-order language** $\mathcal{L}$ is an infinite collection of distinct symbols, no one of which is properly contained in another, separated into the following categories:

1. *Parentheses:* (,).
2. *Connectives:* $\vee$, $\neg$.
3. *Quantifier:* $\forall$.
4. *Variables, one for each positive integer n:* $v_1, v_2, \ldots, v_n, \ldots$. The set of variable symbols will be denoted *Vars*.
5. *Equality symbol:* $=$.
6. *Constant symbols:* Some set of zero or more symbols.
7. *Function symbols:* For each positive integer n, some set of zero or more n-ary function symbols.
8. *Relation symbols:* For each positive integer n, some set of zero or more n-ary relation symbols.

To say that a function symbol is n-ary (or has arity n) means that it is intended to represent a function of n variables. For example, $+$ has arity 2. Similarly, an n-ary relation symbol will be intended to represent a relation on n-tuples of objects. This will be made formal in Definition 1.6.1.

To specify a language, all we have to do is determine which, if any, constant, function, and relation symbols we wish to use. Many authors, by the way, let the equality symbol be optional, or treat the equality symbol as an ordinary binary (i.e., 2-ary) relation symbol. We will assume that each language has the equality symbol, unless specifically noted.

> *Chaff:* I ought to add a word about the phrase "no one of which is properly contained in another," which appears in this definition. We have been quite vague about the meaning of the word *symbol*, but you are supposed to be thinking about marks made on a piece of paper. We will be constructing sequences of symbols and trying to figure out what they mean in the next few pages, and by not letting one symbol be contained in another, we will find our job of interpreting sequences to be much easier.
>
> For example, suppose that our language contained both the constant symbol $\heartsuit$ and the constant symbol $\heartsuit\heartsuit$ (notice that the first symbol is properly contained in the second). If you were reading a sequence of symbols and ran across $\heartsuit\heartsuit$, it would be impossible to decide if this was one symbol or a sequence of two symbols. By not allowing symbols to be contained in other symbols, this type of confusion is avoided, leaving the field open for other types of confusion to take its place.

Example 1.2.2. Suppose that we were taking an abstract algebra course and we wanted to specify the language of groups. A group consists of a set and a binary operation that has certain properties. Among those properties is the existence of an identity element for the operation. Thus, we could decide that our language will contain one constant symbol for the identity element, one binary operation symbol, and no relation symbols. We would get

$$\mathcal{L}_G \text{ is } \{0, +\},$$

where 0 is the constant symbol and + is a binary function symbol. Or perhaps we would like to write our groups using the operation as multiplication. Then a reasonable choice could be

$$\mathcal{L}_G \text{ is } \{1, ^{-1}, \cdot\},$$

which includes not only the constant symbol 1 and the binary function symbol $\cdot$, but also a unary (or 1-ary) function symbol $^{-1}$, which is designed to pick out the inverse of an element of the group. As you can see, there is a fair bit of choice involved in designing a language.

Example 1.2.3. The language of set theory is not very complicated at all. We will include one binary relation symbol, $\in$, and that is all:

$$\mathcal{L}_{ST} \text{ is } \{\in\}.$$

The idea is that this symbol will be used to represent the elementhood relation, so the interpretation of the string $x \in y$ will be that the set x is an element of the set y. You might be tempted to add other relation symbols, such as $\subset$, or constant symbols, such as $\emptyset$, but it will be easier to define such symbols in terms of more primitive symbols. Not easier in terms of readability, but easier in terms of proving things about the language.

In general, to specify a language we need to list the constant symbols, the function symbols, and the relation symbols. There can be infinitely many [in fact, uncountably many (cf. the Appendix)] of each. So, here is a specification of a language:

$$\mathcal{L} \text{ is } \{c_1, c_2, \ldots, f_1^{a(f_1)}, f_2^{a(f_2)}, \ldots, R_1^{a(R_1)}, R_2^{a(R_2)}, \ldots\}.$$

Here, the c_i's are the constant symbols, the $f_i^{a(f_i)}$'s are the function symbols, and the $R_i^{a(R_i)}$'s are the relation symbols. The superscripts on the function and relation symbols indicate the arity of the associated symbols, so a is a mapping that assigns a natural number to a string that begins with an f or an R, followed by a subscripted ordinal. Thus, an official function symbol might look like this:

$$f_{17}^{223},$$

which would say that the function that will be associated with the 17th function symbol is a function of 223 variables. Fortunately, such dreadful detail will rarely be needed. We will usually see only unary or binary function symbols and the arity of each symbol will be stated once. Then the author will trust that the context will remind the patient reader of each symbol's arity.

1.2.1 Exercises

1. Carefully write out the symbols that you would want to have in a language $\mathcal{L}$ that you intend to use to write statements of elementary algebra. Indicate which of the symbols are constant symbols, and the arity of the function and relation symbols that you choose. Now write out another language, $\mathcal{M}$ (i.e., another list of symbols) with the same number of constant symbols, function symbols, and relation symbols that you would *not* want to use for elementary algebra. Think about the value of good notation.

2. What are good examples of unary (1-ary) functions? Binary functions? Can you find natural examples of relations with arity 1, 2, 3, and 4? As you think about this problem, stay mindful of the difference between the function and the function symbol, between the relation and the relation symbol.

3. In the town of Sneezblatt there are three eating establishments: McBurgers, Chez Fancy, and Sven's Tandoori Palace. Think for a minute about statements that you might want to make about these restaurants, and then write out $\mathcal{L}$, the formal language for your theory of restaurants. Have fun with this, but try to include both function and relation symbols in $\mathcal{L}$. What interpretations are you planning for your symbols?

4. You have been put in charge of drawing up the schedule for a basketball league. This league involves eight teams, each of which must play each of the other seven teams exactly two times: once at home and once on the road. Think of a reasonable language for this situation. What constants would you need? Do you need any relation symbols? Function symbols? It would be nice if your finished schedule did not have any team playing two games on the same day. Can you think of a way to state this using the formal symbols that you have chosen? Can you express the sentence which states that each team plays every other team exactly two times?

5. Let's work out a language for elementary trigonometry. To get you started, let me suggest that you start off with *lots* of constant symbols—one for each real number. It is tempting to use the symbol 7 to stand for the number seven, but this runs into problems. (Do you see why this is illegal? 7, 77, 7/3, ...) Now,

what functions would you like to discuss? Think of symbols for them. What are the arities of your function symbols? Do not forget that you need symbols for addition and multiplication! What relation symbols would you like to use?

6. A computer language is another example of a language. For example, the symbol := might be a binary function symbol, where the interpretation of the instruction

$$x := 7$$

would be to alter the internal state of the computer by placing the value 7 into the position in memory referenced by the variable x. Think about the function associated with the binary function symbol

if ______, then ______.

What are the inputs into this function? What sort of thing does the function do? Look at the statement

If $x + y > 3$, then $z := 7$.

Identify the function symbols, constant symbols, and relation symbols. What are the arities of each function and relation symbol?

7. What would be a good language for the theory of vector spaces? This problem is slightly more difficult, as there are two different varieties of objects, scalars and vectors, and you have to be able to tell them apart. Write out the axioms of vector spaces in your language. Or, better yet, use a language that includes a unary function symbol for each real number so that scalars don't exist as objects at all!

8. It is not actually necessary to include function symbols in the language, since a function is just a special kind of relation. Just to see an example, think about the function $f : \mathbb{N} \to \mathbb{N}$ defined by $f(x) = x^2$. Remembering that a relation on $\mathbb{N} \times \mathbb{N}$ is just a set of ordered pairs of natural numbers, find a relation R on $\mathbb{N} \times \mathbb{N}$ such that (x, y) is an element of R if and only if $y = f(x)$. Convince yourself that you could do the same for any function defined on any domain. What condition must be true if a relation R on $A \times B$ is to be a function mapping A to B?

1.3 Terms and Formulas

Suppose that $\mathcal{L}$ is the language $\{0, +, <\}$, and we are going to use $\mathcal{L}$ to discuss portions of arithmetic. If I were to write down the string of symbols from $\mathcal{L}$,

$$(v_1 + 0) < v_1,$$

and the string

$$v_{17})(\forall + +(((0,$$

you would probably agree that the first string conveyed some meaning, even if that meaning were incorrect, while the second string was meaningless. It is our goal in this section to carefully define which strings of symbols of $\mathcal{L}$ we will use. In other words, we will select the strings that will have meaning.

Now, the point of having a language is to be able to make statements about certain kinds of mathematical systems. Thus, we will want the statements in our language to have the ability to refer to objects in the mathematical structures under consideration. So we will need some of the strings in our language to refer to those objects. Those strings are called the terms of $\mathcal{L}$.

Definition 1.3.1. If $\mathcal{L}$ is a language, a **term of** $\mathcal{L}$ is a nonempty finite string t of symbols from $\mathcal{L}$ such that either:

1. t is a variable, or
2. t is a constant symbol, or
3. t is $ft_1t_2 \dots t_n$, where f is an n-ary function symbol of $\mathcal{L}$ and each of the t_i is a term of $\mathcal{L}$.

This is a definition by recursion, as you notice that in the third clause of the definition, t is a term if it contains substrings that are terms. Since the substrings of t are shorter (contain fewer symbols) than t, and as none of the symbols of $\mathcal{L}$ are made up of other symbols of $\mathcal{L}$, this causes no problems.

Example 1.3.2. Let $\mathcal{L}$ be the language $\{\overline{0}, \overline{1}, \overline{2}, \dots, +, \cdot\}$, with one constant symbol for each natural number and two binary function symbols. Here are some of the terms of $\mathcal{L}$: $\overline{714}$, $+\overline{3}\,\overline{2}$, $\cdot + \overline{3}\,\overline{2}\,\overline{4}$. Notice that $\overline{1}\,\overline{2}\,\overline{3}$ is not a term of $\mathcal{L}$, but rather is a sequence of three terms in a row.

> *Chaff:* The term $+\overline{3}\,\overline{2}$ looks pretty annoying at this point, but we will use this sort of notation (called *Polish notation*) for functions rather than the infix notation ($\overline{3}+\overline{2}$) that you are used to. We are not really being that odd here: You have certainly seen some functions written in Polish notation: $\sin(x)$ and $f(x, y, z)$ come to mind. We are just being consistent in treating addition in the same way. What makes it difficult is that it is hard to remember that addition really is just another function of two variables. But I am sure that by the end of this book, you will be very comfortable with that idea and with the notation that we are using.

A couple of points are probably worth emphasizing, just this once. Notice that in the application of the function symbols, there are no parentheses and no commas. Also notice that all of our functions are written with the operator on the left. So instead of $\overline{3}+\overline{2}$, we write $+\overline{3}\,\overline{2}$. The reason for this is for consistency and to make sure that we can parse our expressions.

Let me give an example. Suppose that, in some language or other, we wrote down the string of symbols $\heartsuit$¥ $\uparrow \diamondsuit\#\# \int$. Assume that two of our colleagues, Humphrey and Ingrid, were waiting in the hall while we wrote down the string. If Humphrey came into the room and announced that our string was a 3-ary function symbol followed by three terms, whereas Ingrid proclaimed that the string was really a 4-ary relation symbol followed by two terms, this would be rather confusing. It would be *really* confusing if they were both correct! So we need to make sure that the strings that we write down can be interpreted in only one way. This property, called *unique readability*, is addressed in Exercise 7 of Section 1.4.1.

> *Chaff:* Unique readability is one of those things that, in the opinion of the author, is important to know, interesting to prove, and boring to read. Thus the proof is placed in (I do not mean "relegated to") the exercises.

Suppose that we look more carefully at the term $\cdot + \overline{3}\,\overline{2}\,\overline{4}$. Assume for now that the symbols in this term are supposed to be interpreted in the usual way, so that $\cdot$ means multiply, $+$ means add, and $\overline{3}$ means three. Then if we add some parentheses to the term in order to clarify its meaning, we get

$$\cdot(+\overline{3}\,\overline{2})\,\overline{4},$$

which ought to have the same meaning as $\cdot\overline{5}\,\overline{4}$, which is $\overline{20}$, just as you suspected.

Rest assured that we will continue to use infix notation, commas, and parentheses as seem warranted to increase the readability (by humans) of this text. So $ft_1t_2\ldots t_n$ will be written $f(t_1, t_2, \ldots, t_n)$ and $+\overline{3}\,\overline{2}$ will be written $\overline{3} + \overline{2}$, with the understanding that this is shorthand and that our official version is the version given in Definition 1.3.1.

The terms of $\mathcal{L}$ play the role of the nouns of the language. To make meaningful mathematical statements about some mathematical structure, we will want to be able to make assertions about the objects of the structure. These assertions will be the formulas of $\mathcal{L}$.

Definition 1.3.3. If $\mathcal{L}$ is a first-order language, a **formula of $\mathcal{L}$** is a nonempty finite string ϕ of symbols from $\mathcal{L}$ such that either:

1. ϕ is $= t_1t_2$, where t_1 and t_2 are terms of $\mathcal{L}$, or
2. ϕ is $Rt_1t_2\ldots t_n$, where R is an n-ary relation symbol of $\mathcal{L}$ and t_1, t_2, ..., t_n are all terms of $\mathcal{L}$, or
3. ϕ is $(\neg\alpha)$, where α is a formula of $\mathcal{L}$, or
4. ϕ is $(\alpha \vee \beta)$, where α and β are formulas of $\mathcal{L}$, or
5. ϕ is $(\forall v)(\alpha)$, where v is a variable and α is a formula of $\mathcal{L}$.

If a formula ψ contains the subformula $(\forall v)(\alpha)$ [meaning that the string of symbols that constitute the formula $(\forall v)(\alpha)$ is a substring of the string of symbols that make up ψ], we will say that the **scope** of the quantifier $\forall$ is α. Any symbol in α will be said to lie within the scope of the quantifier $\forall$. Notice that a formula ψ can have several different occurrences of the symbol $\forall$, and each occurrence of the quantifier will have its own scope. Also notice that one quantifier can lie within the scope of another.

The **atomic formulas of $\mathcal{L}$** are those formulas that satisfy clause (1) or (2) of Definition 1.3.3.

You have undoubtedly noticed that there are no parentheses or commas in the atomic formulas, and you have probably decided that we will continue to use both commas and infix notation as seems appropriate. You are correct on both counts. So, instead of writing the official version

$$< SSSSS0SS0$$

in a language containing constant symbol 0, unary function symbol S, and binary relation symbol $<$, we will write

$$SSSSS0 \quad < \quad SS0$$

or (after some preliminary definitions)

$$\overline{5} < \overline{2}.$$

Also notice that we *are* using infix notation for the binary logical connective $\lor$. I hope that this will make your life somewhat easier.

You will be asked in Exercise 8 in Section 1.4.1 to prove that unique readability holds for formulas as well as terms. We will, in our exposition, use different-size parentheses, different shapes of delimiters, and omit parentheses in order to improve readability without (we hope) introducing confusion on your part.

Notice that a term is not a formula! If the terms are the nouns of the language, the formulas will be the statements. Statements can be either true or false. Nouns cannot. Much confusion can be avoided if you keep this simple dictum in mind.

For example, suppose that you are looking at a string of symbols and you notice that the string does not contain either the symbol $=$ or any other relation symbol from the language. Such a string cannot be a formula, as it makes no claim that can be true or false. The string might be a term, it might be nonsense, but it cannot be a formula.

> *Chaff:* I do hope that you have noticed that we are dealing only with the syntax of our language here. We have not mentioned that the symbol $\neg$ will be used for denial, or that $\lor$ will mean "or," or even that $\forall$ means "for every." Don't worry, they will mean what you think they should mean. Similarly, do not worry about the fact that the definition of a formula left out symbols for conjunctions, implications, and biconditionals. We will get to them in good time.

1.3.1 Exercises

1. Suppose that the language $\mathcal{L}$ consists of two constant symbols, $\diamondsuit$ and $\heartsuit$, a unary relation symbol ¥, a binary function symbol $\flat$, and a 3-ary function symbol $\sharp$. Write down at least three distinct terms of the language $\mathcal{L}$. Write down a couple of nonterms that look like they might be terms and explain why they are not terms. Write a couple of formulas and a couple of nonformulas that look like they ought to be formulas.

2. The fact that we write all of our operations on the left is important for unique readability. Suppose, for example, that we wrote our binary operations in the middle (and did not allow the use of parentheses). If our language included the binary function symbol #, then the term

$$u\#v\#w$$

could be interpreted two ways. This can make a difference: Suppose that the operation associated with the function symbol # is "subtract." Find three real numbers u, v, and w such that the two different interpretations of $u\#v\#w$ lead to different answers. Any nonassociative binary function will yield another counterexample to unique readability. Can you think of three such functions?

3. The language of number theory is

$$\mathcal{L}_{NT} \text{ is } \{0, S, +, \cdot, E, <\},$$

where the intended meanings of the symbols are as follows: 0 stands for the number zero, S is the successor function $S(x) = x + 1$, the symbols $+$, $\cdot$, and $<$ mean what you expect, and E stands for exponentiation, so $E(3, 2) = 9$. Assume that $\mathcal{L}_{NT}$-formulas will be interpreted with respect to the nonnegative integers and write an $\mathcal{L}_{NT}$-formula to express the claim that p is a prime number. Can you write a formula stating that there is no largest prime number? Can you write the statement of Lagrange's Theorem, which states that every natural number is the sum of four squares? What is the formal statement of the Twin Primes Conjecture, which says that there are infinitely many pairs (x, y) such that x and y are both prime and $y = x + 2$?

Use shorthand in your answer to this problem. After you have found the formula which says that p is prime, call the formula $Prime(p)$, and use $Prime(p)$ in your later answers.

4. Suppose that our language has infinitely many constant symbols of the form $'$, $''$, $'''$, ... and no function or relation symbols other than $=$. Explain why this situation leads to problems by looking at the formula $=''''''$. Where in our definitions do we outlaw this sort of problem?

1.4 Induction

You are familiar, no doubt, with proofs by induction. They are the bane of most mathematics students from their first introduction in high school through the college years. It is my goal in this section to discuss the proofs by induction that you know so well, put them in a different light, and then generalize that notion of induction to a setting that will allow us to use induction to prove things about terms and formulas rather than just the natural numbers.

Just to remind you of the general form of a proof by induction on the natural numbers, let me state and prove a familiar theorem, assuming for the moment that the set of natural numbers is $\{1, 2, 3, \ldots\}$.

Theorem 1.4.1. *For every natural number n,*

$$1 + 2 + \cdots + n = \frac{n(n+1)}{2}.$$

Proof: If $n = 1$, simple computation shows that the equality holds. For the inductive case, fix $k \geq 1$ and assume that

$$1 + 2 + \cdots + k = \frac{k(k+1)}{2}.$$

If we add $k + 1$ to both sides of this equation, we get

$$1 + 2 + \cdots + k + (k+1) = \frac{k(k+1)}{2} + (k+1),$$

and simplifying the right-hand side of this equation shows that

$$1 + 2 + \cdots + (k+1) = \frac{(k+1)\big[(k+1)\big((k+1)+1\big)\big]}{2},$$

finishing the inductive step, and the proof. ■

As you look at the proof of this theorem, you notice that there is a base case, when $n = 1$, and an inductive case. In the inductive step of the proof, we prove the implication

If the formula holds for k, then the formula holds for $k + 1$.

We prove this implication by assuming the antecedent, that the theorem holds for a (fixed, but unknown) number k, and from that assumption proving the consequent, that the theorem holds for the next number, $k + 1$. Notice that this is *not* the same as assuming the theorem that we are trying to prove. The theorem is a universal statement—it claims that a certain formula holds for every natural number.

Looking at this from a slightly different angle, what we have done is to construct a set of numbers with a certain property. If we let S stand for the set of numbers for which our theorem holds, in our proof by induction we show the following facts about S:

1. The number 1 is an element of S. We prove this explicitly in the base case of the proof.

2. If the number k is an element of S, then the number $k + 1$ is an element of S. This is the content of the inductive step of the proof.

But now, notice that we know that the collection of natural numbers can be defined as the smallest set such that:

1. The number 1 is a natural number.

2. If k is a natural number, then $k + 1$ is a natural number.

So S, the collection of numbers for which the theorem holds, is identical with the set of natural numbers, thus the theorem holds for every natural number n, as needed. (If you caught the slight lie here, just substitute "superset" where appropriate.)

So what makes a proof by induction work is the fact that the natural numbers can be defined recursively. There is a base case, consisting of the smallest natural number ("1 is a natural number"), and there is a recursive case, showing how to construct bigger natural numbers from smaller ones ("If k is a natural number, then $k + 1$ is a natural number").

Now, let us look at Definition 1.3.3, the definition of a formula. Notice that the five clauses of the definition can be separated into two groups. The first two clauses, the atomic formulas, are explicitly defined: For example, the first case says that anything that is of the form $= t_1t_2$ is a formula if t_1 and t_2 are terms. These first two clauses form the base case of the definition. The last three clauses are the recursive case, showing how if α and β are formulas, they can be used to build more complex formulas, such as $(\alpha \vee \beta)$ or $(\forall v)(\alpha)$.

Now since the collection of formulas is defined recursively, we can use an inductive-style proof when we want to prove that something is true about *every* formula. The inductive proof will consist of two parts, a base case and an inductive case. In the base case of the proof we will verify that the theorem is true about every atomic formula—about every string that is known to be a formula from the base case of the definition. In the inductive step of the proof, we assume that the theorem is true about simple formulas (α and β), and use that assumption to prove that the theorem holds a more complicated formula ϕ that is generated by a recursive clause of the definition. This method of proof is called *induction on the complexity of the formula.*

The following theorem, although mildly interesting in its own right, is included here mostly so that the reader can see an example of a proof by induction in this setting:

Theorem 1.4.2. *Suppose that ϕ is a formula in the language $\mathcal{L}$. Then the number of left parentheses occurring in ϕ is equal to the number of right parentheses occurring in ϕ.*

Proof: We will present this proof in a fair bit of detail, in order to emphasize the proof technique. As you become accustomed to proving theorems by induction on complexity, not so much detail is needed.

Base Case. We begin our inductive proof with the base case, as you would expect. Our theorem makes an assertion about all formulas, and the simplest formulas are the atomic formulas. They constitute our base case. Suppose that ϕ is an atomic formula. There are two varieties of atomic formulas: Either ϕ begins with an equals sign followed by two terms, or ϕ begins with a relation symbol followed by several terms. As there are no parentheses in any term (we are using the official definition of term, here), there are no parentheses in ϕ. Thus, there are as many left parentheses as right parentheses in ϕ, and we have established the theorem if ϕ is an atomic formula.

Inductive Case. The inductive step of a proof by induction on complexity of a formula takes the following form: Assume that ϕ is a formula by virtue of clause (3), (4), or (5) of Definition 1.3.3. Also assume that the statement of the theorem is true when applied to the formulas α and β. With those assumptions we will prove that the statement of the theorem is true when applied to the formula ϕ. Thus, as every formula is a formula either by virtue of being an atomic formula or by application of clause (3), (4), or (5) of the definition, we will have shown that the statement of the theorem is true when applied to any formula, which has been our goal.

So, assume that α and β are formulas that contain equal numbers of left and right parentheses. Suppose that there are k left parentheses and k right parentheses in α and l left parentheses and l right parentheses in β.

If ϕ is a formula by virtue of clause (3) of the definition, then ϕ is $(\neg\alpha)$. We observe that there are $k+1$ left parentheses and $k+1$ right parentheses in ϕ, and thus ϕ has an equal number of left and right parentheses, as needed.

If ϕ is a formula because of clause (4), then ϕ is $(\alpha \lor \beta)$, and ϕ contains $k+l+1$ left and right parentheses, an equal number of each type.

Finally, if ϕ is $(\forall v)(\alpha)$, then ϕ contains $k+2$ left parentheses and $k+2$ right parentheses, as needed.

This concludes the possibilities for the inductive case of the proof, so we have established that in every formula, the number of left parentheses is equal to the number of right parentheses. ■

This theorem is a little cheesy, but the power of this proof technique will become more evident as you work through the following exercises and when we discuss the semantics of our language.

Notice also that the definition of a term (Definition 1.3.1) is also a recursive definition, so we can use induction on the complexity of a term to prove that a theorem holds for every term.

1.4.1 Exercises

1. Prove, by ordinary induction on the natural numbers, that

$$1^2 + 2^2 + \cdots + n^2 = \frac{n(n+1)(2n+1)}{6}.$$

2. Prove, by induction, that the sum of the interior angles in a convex n-gon is $(n-2)180^\circ$. (A convex n-gon is a polygon with n sides, where the interior angles are all less than 180°.)

3. Prove by induction that if A is a set consisting of n elements, then A has 2^n subsets.

4. Suppose that $\mathcal{L}$ is $\{0, f, g\}$, where 0 is a constant symbol, f is a binary function symbol, and g is a 4-ary function symbol. Use induction on complexity to show that every $\mathcal{L}$-term has an odd number of symbols.

5. If $\mathcal{L}$ is $\{0, <\}$, where 0 is a constant symbol and $<$ is a binary relation symbol, show that the number of symbols in any formula is divisible by 3.

6. If s and t are strings, we say that s is an *initial segment* of t if there is a nonempty string u such that $t = su$, where su is the string s followed by the string u. For example, KUMQ is an initial segment of KUMQUAT and $+24$ is an initial segment of $+24u - v$. Prove, by induction on the complexity of s, that if s and t are terms, then s is not an initial segment of t. [*Suggestion:* The base case, when s is either a variable or a constant symbol, should be easy. Then suppose that s is an initial segment of t and $s = ft_1t_2 \ldots t_n$, where you know that each t_i is not an initial segment of any other term. Look for a contradiction.]

7. A language is said to satisfy unique readability for terms if, for each term t, t is in exactly one of the following categories:

   (a) Variable

   (b) Constant symbol

   (c) Complex term

   and furthermore, if t is a complex term, then there is a unique function symbol f and a unique sequence of terms $t_1, t_2, \ldots, t_n$

such that $t = ft_1t_2 \ldots t_n$. Prove that our languages satisfy unique readability for terms. [*Suggestion:* You mostly have to worry about uniqueness—for example, suppose that t is c, a constant symbol. How do you know that t is not also a complex term? Suppose that t is $ft_1t_2 \ldots t_n$. How do you show that the f and the t_i's are unique? You may find Exercise 6 useful.]

8. To say that a language satisfies unique readability for formulas is to say that every formula ϕ is in exactly one of the following categories:

(a) Equality (if ϕ is $= t_1t_2$)

(b) Other atomic (if ϕ is $Rt_1t_2 \ldots t_n$ for an n-ary relation symbol R)

(c) Negation

(d) Disjunction

(e) Quantified

Also, it must be that if ϕ is both $= t_1t_2$ and $= t_3t_4$, then t_1 is identical to t_3 and t_2 is identical to t_4, and similarly for other atomic formulas. Furthermore, if (for example) ϕ is a negation $(\neg\alpha)$, then it must be the case that there is not another formula β such that ϕ is also $(\neg\beta)$, and similarly for disjunctions and quantified formulas. Prove that our languages satisfy unique readability for formulas. You will want to look at, and use, Exercise 7. You may have to prove an analog of Exercise 6, in which it may be helpful to think about the parentheses in an initial segment of a formula, in order to prove that no formula is an initial segment of another formula.

9. Take the proof of Theorem 1.4.2 and write it out in the way that you would present it as part of a homework assignment. Thus, you should cut out all of the inessential motivation and present only what is needed to make the proof work.

1.5 Sentences

Among the formulas in the language $\mathcal{L}$, there are some in which we will be especially interested. These are the sentences of $\mathcal{L}$—the formulas that can be either true or false in a given mathematical model.

Let me use an example to introduce a language that will be vitally important to us as we work through this book.

Definition 1.5.1. The language $\mathcal{L}_{NT}$ is $\{0, S, +, \cdot, E, <\}$, where 0 is a constant symbol, S is a unary function symbol, $+$, $\cdot$, and E are binary function symbols, and $<$ is a binary relation symbol. This will be referred to as the language of number theory.

> *Chaff:* Although we are not fixing the meanings of these symbols yet, I probably ought to tell you that the standard interpretation of $\mathcal{L}_{NT}$ will use 0, $+$, $\cdot$, and $<$ in the way that you expect. The symbol S will stand for the successor function that maps a number x to the number $x+1$, and E will be used for exponentiation: $E32$ is supposed to be 3^2.

Consider the following two formulas of $\mathcal{L}_{NT}$:

$$\neg(\forall x)[(y < x) \lor (y = x)].$$
$$(\forall x)(\forall y)[(x < y) \lor (x = y) \lor (y < x)].$$

(Did you notice that we have begun using an informal presentation of the formulas?)

The second formula should look familiar. It is nothing more than the familiar trichotomy law of $<$, and you would agree that the second formula is a true statement about the collection of natural numbers, where you are interpreting $<$ in the usual way.

The first formula above is different. It "says" that not every x is greater than or equal to y. The truth of that statement is indeterminate: It depends on what natural number y represents. The formula might be true, or it might be false—it all depends on the value of y. So our goal in this section is to separate the formulas of $\mathcal{L}$ into one of two classes: the sentences (like the second example above) and the nonsentences. To begin this task, we must talk about free variables.

Free variables are the variables upon which the truth value of a formula may depend. The variable y is free in the first formula above. To draw an analogy from calculus, if we look at

$$\int_1^x \frac{1}{t}\,dt,$$

the variable x is free in this expression, as the value of the integral depends on the value of x. The variable t is not free, and in fact it doesn't make any sense to decide on a value for t. The same distinction holds between free and nonfree variables in an $\mathcal{L}$-formula. Let us try to make things a little more precise.

Definition 1.5.2. Suppose that v is a variable and ϕ is a formula. We will say that v **is free in** ϕ if

1. ϕ is atomic and v occurs in (is a symbol in) ϕ, or
2. ϕ is $(\neg\alpha)$ and v is free in α, or
3. ϕ is $(\alpha \vee \beta)$ and v is free in at least one of α or β, or
4. ϕ is $(\forall u)(\alpha)$ and v is not u and v is free in α.

Thus, if we look at the formula

$$\forall v_2 \neg(\forall v_3)(v_1 = S(v_2) \vee v_3 = v_2),$$

the variable v_1 is free whereas the variables v_2 and v_3 are not free. A slightly more complicated example is

$$(\forall v_1 \forall v_2(v_1 + v_2 = 0)) \vee v_1 = S(0).$$

In this formula, v_1 is free whereas v_2 is not free. Especially when a formula is presented informally, you must be careful about the scope of the quantifiers and the placement of parentheses.

We will have occasion to use the informal notation $\forall x \phi(x)$. This will mean that ϕ is a formula and x is the only free variable of ϕ. If we then write $\phi(t)$, where t is an $\mathcal{L}$-term, that will denote the formula obtained by taking ϕ and replacing each occurrence of the variable x with the term t. This will all be defined more formally and more precisely in Definition 1.8.2.

Definition 1.5.3. A **sentence** in a language $\mathcal{L}$ is a formula of $\mathcal{L}$ that contains no free variables.

For example, if a language contained the constant symbols 0, 1, and 2 and the binary function symbol $+$, then the following are sentences: $1+1 = 2$ and $(\forall x)(x+1 = x)$. You are probably convinced that the first of these is true and the second of these is false. In the next two sections we will see that you might be correct. But then again, you might not be.

1.5.1 Exercises

1. For each of the following, find the free variables, if any, and decide if the given formula is a sentence. The language includes a binary function symbol $+$, a binary relation symbol $<$, and constant symbols 0 and 2.

(a) $(\forall x)(\forall y)(x + y = 2)$

(b) $(x + y < x) \vee (\forall z)(z < 0)$

(c) $((\forall y)(y < x)) \vee ((\forall x)(x < y))$

2. Explain precisely, using the definition of a free variable, how you know that the variable v_2 is free in the formula

$$(\forall v_1)(\neg(\forall v_5)(v_2 = v_1 + v_5)).$$

3. In mathematics, we often see statements such as $\sin^2 x + \cos^2 x = 1$. Notice that this is not a sentence, as the variable x is free. But we all agree that this statement is true, given the usual interpretations of the symbols. How can we square this with the claim that *sentences* are the formulas that can be either true or false?

4. If we look at the first of our example formulas in this section,

$$\neg(\forall x)[(y < x) \vee (y = x)],$$

and we interpret the variables as ranging over the natural numbers, you will probably agree that the formula is false if y represents the natural number 0 and true if y represents any other number. (If you aren't happy with 0 being a natural number, then use 1.) On the other hand, if we interpret the variables as ranging over the integers, what can we say about the truth or falsehood of this formula? Can you think of an interpretation for the symbols that would make sense if we try to apply this formula to the collection of complex numbers?

5. A variable may occur several times in a given formula. For example, the variable v_1 occurs four times in the formula

$$(\forall v_1)\big[(v_1 = v_3) \vee (v_1 = Sv_2) \vee (0 + v_{17} < v_1 - S0)\big].$$

What should it mean for an *occurrence* of a variable to be free? Write a definition that begins: The nth occurrence of a variable v in a formula ϕ is said to be free if .... An occurrence of v in ϕ that is not free is said to be **bound**. Give an example of a formula in a suitable language that contains both free and bound occurrences of a variable v.

6. Look at the formula

$$\left[(\forall y)(x = y)\right] \vee \left[(\forall x)(x < 0)\right].$$

If we denote this formula by $\phi(x)$ and t is the term $S0$, find $\phi(t)$. [*Suggestion:* The trick here is to see that there is a bit of a lie in the discussion of $\phi(t)$ in the text. Having completed Exercise 5, we can now say that we only replace the free occurrences of the variable x when we move from $\phi(x)$ to $\phi(t)$.]

1.6 Structures

Let us, by way of example, return to the language $\mathcal{L}_{NT}$ of number theory. Recall that $\mathcal{L}_{NT}$ is $\{0, S, +, \cdot, E, <\}$, where 0 is a constant symbol, S is a unary function symbol, $+$, $\cdot$, and E are binary function symbols, and $<$ is a binary relation symbol. We now want to discuss the possible mathematical structures in which we can interpret these symbols, and thus the formulas and sentences of $\mathcal{L}_{NT}$.

"But wait!" cries the incredulous reader. "You just said that this is the language of number theory, so certainly we already know what each of those symbols means."

It is certainly the case that you know *an* interpretation for these symbols. The point of this section is that there are *many* different possible interpretations for these symbols, and we want to be able to specify which of those interpretations we have in mind at any particular moment.

Probably the interpretation you had in mind (what we will call the standard model for number theory) works with the set of natural numbers $\{0, 1, 2, 3, \dots\}$. The symbol 0 stands for the number 0.

> *Chaff:* Carefully, now! The symbol 0 is the mark on the paper, the numeral. The number 0 is the thing that the numeral 0 represents. The numeral is something that

you can see. The number is something that you cannot see.

The symbol S is a unary function symbol, and the function for which that symbol stands is the successor function that maps a number to the next larger natural number. The symbols $+$, $\cdot$, and E represent the functions of addition, multiplication, and exponentiation, and the symbol $<$ will be used for the "less than" relation.

But that is only one of the ways that we might choose to interpret those symbols. Another way to interpret all of those symbols would be to work with the numbers 0 and 1, interpreting the symbol 0 as the number 0, S as the function that maps 0 to 1 and 1 to 0, $+$ as addition mod 2, $\cdot$ as multiplication mod 2, and (just for variety) E as the function with constant value 1. The symbol $<$ can still stand for the relation "less than."

Or, if I were in a slightly more bizarre mood, I could work in a universe consisting of Beethoven, Picasso, and Ernie Banks, interpreting the symbol 0 as Picasso, S as the identity function, $<$ as equality, and each of the binary function symbols as the constant function with output Ernie Banks.

The point is that there is nothing sacred about one mathematical structure as opposed to another. Without determining the structure under consideration, without deciding how we wish to interpret the symbols of the language, we have no way of talking about the truth or falsity of a sentence as trivial as

$$(\forall v_1)(v_1 < S(v_1)).$$

Definition 1.6.1. Fix a language $\mathcal{L}$. An **$\mathcal{L}$-structure** $\mathfrak{A}$ is a nonempty set A, called the **universe of** $\mathfrak{A}$, together with:

1. For each constant symbol c of $\mathcal{L}$, an element $c^{\mathfrak{A}}$ of A
2. For each n-ary function symbol f of $\mathcal{L}$, a function $f^{\mathfrak{A}} : A^n \to A$
3. For each n-ary relation symbol R of $\mathcal{L}$, an n-ary relation $R^{\mathfrak{A}}$ on A (i.e., a subset of A^n)

Chaff: The letter $\mathfrak{A}$ is a German Fraktur capital A. We will also have occasion to use $\mathfrak{A}$'s friends, $\mathfrak{B}$ and $\mathfrak{C}$. $\mathfrak{N}$ will be used for a particular structure involving the natural numbers. The use of this typeface is traditional (which means this is the way I learned it). For your handwritten work, probably using capital script letters will be the best.

Often, we will write a structure as an ordered k-tuple, like this:

$$\mathfrak{A} = \langle A, c_1^{\mathfrak{A}}, c_2^{\mathfrak{A}}, f_1^{\mathfrak{A}}, R_1^{\mathfrak{A}}, R_2^{\mathfrak{A}} \rangle.$$

As you can see, the notation is starting to get out of hand once again, and we will not hesitate to simplify and abbreviate when we believe that we can do so without confusion. So, when we are working in $\mathcal{L}_{NT}$, we will often talk about the standard structure

$$\mathfrak{N} = \langle \mathbb{N}, 0, S, +, \cdot, E, < \rangle,$$

where the constants, functions, and relations do not get the superscripts they deserve, and the author trusts you will interpret $\mathbb{N}$ as the collection $\{0, 1, 2, \dots\}$ of natural numbers, the symbol 0 to stand for the number zero, $+$ to stand for addition, S to stand for the successor function, and so on. By the way, if you are not used to thinking of 0 as a natural number, do not panic. Set theorists see 0 as the most natural of objects, so we tend to include it in $\mathbb{N}$ without thinking about it.

Example 1.6.2. The structure $\mathfrak{N}$ that we have just introduced is called the standard $\mathcal{L}_{NT}$-structure. To emphasize that there are other perfectly good $\mathcal{L}_{NT}$-structures, let us construct a different $\mathcal{L}_{NT}$-structure $\mathfrak{A}$ with exactly four elements. The elements of A will be Oberon, Titania, Puck, and Bottom. The constant $0^{\mathfrak{A}}$ will be Bottom. Now we have to construct the functions and relations for our structure. As everything is unary or binary, setting forth tables (as in Table 1.1) seems a reasonable way to proceed. So you can see that in this structure $\mathfrak{A}$ that Titania + Puck = Oberon, while Puck + Titania = Titania. You can also see that 0 (also known as Bottom) is not the additive identity in this structure, and that $<$ is a very strange ordering.

Now the particular functions and relation that I chose were just the functions and relations that jumped into my fingers as I typed up this example, but any such functions would have worked perfectly well to define an $\mathcal{L}_{NT}$-structure. It may well be worth your while to figure out if this $\mathcal{L}_{NT}$-sentence is true (whatever that means) in $\mathfrak{A}$: $SS0 + SS0 < SSSSS0E0 + S0$.

x	$S^{\mathfrak{A}}(x)$
Oberon	Oberon
Titania	Bottom
Puck	Titania
Bottom	Titania

$+^{\mathfrak{A}}$	Oberon	Titania	Puck	Bottom
Oberon	Puck	Puck	Puck	Titania
Titania	Puck	Bottom	Oberon	Titania
Puck	Bottom	Titania	Bottom	Titania
Bottom	Bottom	Bottom	Bottom	Oberon

$\cdot^{\mathfrak{A}}$	Oberon	Titania	Puck	Bottom
Oberon	Oberon	Titania	Puck	Bottom
Titania	Titania	Bottom	Oberon	Titania
Puck	Bottom	Bottom	Oberon	Oberon
Bottom	Titania	Oberon	Puck	Puck

$E^{\mathfrak{A}}$	Oberon	Titania	Puck	Bottom
Oberon	Puck	Puck	Oberon	Oberon
Titania	Titania	Titania	Titania	Titania
Puck	Titania	Bottom	Oberon	Puck
Bottom	Bottom	Puck	Titania	Puck

$<^{\mathfrak{A}}$	Oberon	Titania	Puck	Bottom
Oberon	Yes	No	Yes	Yes
Titania	No	No	Yes	No
Puck	Yes	Yes	Yes	Yes
Bottom	No	No	Yes	No

Table 1.1: A Midsummer Night's Structure

Example 1.6.3. We work in a language with one constant symbol, $\pounds$, and one unary function symbol, X. So, to define a model $\mathfrak{A}$, all we need to do is specify a universe, an element of the universe, and a function $X^{\mathfrak{A}}$. Suppose that we let the universe be the collection of all finite strings of 0 or more capital letters from the Roman alphabet. So A includes such strings as: BABY, LOGICISBETTERTHANSIX, ε (the empty string), and DLKFDFAHADS. The constant symbol $\pounds$ will be interpreted as the string POTITION, and the function $X^{\mathfrak{A}}$ is the function that adds an X to the beginning of a string. So $X^{\mathfrak{A}}(\text{YLOPHONE}) = \text{XYLOPHONE}$. Convince yourself that this is a valid, if somewhat odd, $\mathcal{L}$-structure.

To try to be clear about things, notice that we have X, the function symbol, which is an element of the language $\mathcal{L}$. Then there is X, the string of exactly one capital letter of the Roman alphabet, which is one of the elements of the universe. (Did you notice the change in typeface without my pointing it out? You may have a future in publishing!)

Let us look at one of the terms of the language: $X\pounds$. In our particular $\mathcal{L}$-structure $\mathfrak{A}$ we will interpret this as

$$X^{\mathfrak{A}}(\pounds^{\mathfrak{A}}) = X^{\mathfrak{A}}(\text{POTITION}) = \text{XPOTITION}.$$

In a different structure, $\mathfrak{B}$, it is entirely possible that the interpretation of the term $X\pounds$ will be HUNNY or AARDVARK or $3\pi/17$. Without knowing the structure, without knowing how to interpret the symbols of the language, we cannot begin to know what object is referred to by a term.

> *Chaff:* All of this stuff about interpreting terms in a structure will be made formal in the next section, so don't panic if it doesn't all make sense right now.

What makes this example confusing, as well as important, is that the function symbol is part of the structure for the language and (modulo a superscript and a change in typeface) the function acts on the elements of the structure in the same way that the function symbol is used in creating $\mathcal{L}$-formulas.

Example 1.6.4. Now, let $\mathcal{L}$ be $\{0, f, g, R\}$, where 0 is a constant symbol, f is a unary function symbol, g is a binary function symbol, and R is a 3-ary relation symbol. We define an $\mathcal{L}$-structure $\mathfrak{B}$ as follows: B, the universe, is the set of all variable-free $\mathcal{L}$-terms. The constant $0^{\mathfrak{B}}$ is the term 0. The functions $f^{\mathfrak{B}}$ and $g^{\mathfrak{B}}$ are defined as in Example 1.6.3, so if t and s are elements of B (i.e., variable-free terms), then $f^{\mathfrak{B}}(t)$ is ft and $g^{\mathfrak{B}}(t,s)$ is gts.

Let us look at this in a little more detail. Consider 0, the constant symbol, which is an element of $\mathcal{L}$. Since 0 is a constant symbol, it is a term, so 0 is an element of B, the universe of our structure $\mathfrak{B}$. (Alas, there is no change in typeface to help us out this time.) If we want to see what element of the universe is referred to by the constant symbol 0, we see that $0^{\mathfrak{B}} = 0$, so the term 0 refers to the element of the universe 0.

If we look at another term of the language, say, $f0$, and we try to find the element of the universe that is denoted by this term, we find that it is

$$f^{\mathfrak{B}}(0^{\mathfrak{B}}) = f^{\mathfrak{B}}(0) = f0.$$

So the term $f0$ denotes an element of the universe, and that element of the universe is ... $f0$. This is pretty confusing, but all that is going on is that the elements of the universe *are* the syntactic objects of the language.

This sort of structure is called a *Henkin structure*, after Leon Henkin, who introduced them in his Ph.D. dissertation in 1949. These structures will be crucial in our proof of the Completeness Theorem in Chapter 3. The proof of that theorem will involve the construction of a particular mathematical structure, and the structure that we will build will be a Henkin structure.

To finish building our structure $\mathfrak{B}$, we have to define a relation $R^{\mathfrak{B}}$. As R is a 3-ary relation symbol, $R^{\mathfrak{B}}$ is a subset of B^3. We will arbitrarily define

$$R^{\mathfrak{B}} = \{\langle r,s,t\rangle \in B^3 \mid \text{the number of function symbols in } r \text{ is even}\}.$$

This finishes defining the structure $\mathfrak{B}$. The definition of $R^{\mathfrak{B}}$ given is entirely arbitrary. I invite you to come up with a more interesting or more humorous definition on your own.

1.6.1 Exercises

1. Consider the structure constructed in Example 1.6.2. Find the value of each of the following: $0+0$, $0E0$, $S0 \cdot SS0$. Do you think $0 < 0$ in this structure?

2. Suppose that $\mathcal{L}$ is the language $\{0, +, <\}$. Let's work together to describe an $\mathcal{L}$-structure $\mathfrak{A}$. Let the universe A be the set consisting of all of the natural numbers together with Ingrid Bergman and Humphrey Bogart. You decide on the interpretations of the symbols. What is the value of $5 +$ Ingrid? Is Bogie < 0?

3. Here is a language consisting of one constant symbol, one 3-ary function symbol, and one binary relation symbol: $\mathcal{L}$ is $\{\flat, \sharp, \natural\}$. Describe an $\mathcal{L}$-model that has as its universe $\mathbb{R}$, the set of real numbers. Describe another $\mathcal{L}$-model that has a finite universe.

4. Write a short paragraph explaining the difference between a language and a structure for a language.

5. Suppose that $\mathfrak{A}$ and $\mathfrak{B}$ are two $\mathcal{L}$-structures. We will say that $\mathfrak{A}$ and $\mathfrak{B}$ are **isomorphic** and write $\mathfrak{A} \cong \mathfrak{B}$ if there is a bijection $i : A \to B$ such that for each constant symbol c of $\mathcal{L}$, $i(c^{\mathfrak{A}}) = c^{\mathfrak{B}}$, for each n-ary function symbol f and for each $a_1, \ldots, a_n \in A$, $i(f^{\mathfrak{A}}(a_1, \ldots, a_n)) = f^{\mathfrak{B}}(i(a_1), \ldots, i(a_n))$, and for each n-ary relation symbol R in $\mathcal{L}$, $\langle a_1, \ldots, a_n \rangle \in R^{\mathfrak{A}}$ if and only if $\langle i(a_1), \ldots, i(a_n) \rangle \in R^{\mathfrak{B}}$. The function i is called an **isomorphism**.

   (a) Show that $\cong$ is an equivalence relation. [*Suggestion:* This means that you must show that the relation $\cong$ is reflexive, symmetric, and transitive. To show that $\cong$ is reflexive, you must show that for any structure $\mathfrak{A}$, $\mathfrak{A} \cong \mathfrak{A}$, which means that you must find an isomorphism, a function, mapping A to A that satisfies the conditions above. So the first line of your proof should be, "Consider this function, with domain A and codomain A: $i(x) =$ something brilliant." Then show that your function i is an isomorphism. Then show, if $\mathfrak{A} \cong \mathfrak{B}$, then $\mathfrak{B} \cong \mathfrak{A}$. Then tackle transitivity. In each case, you must define a particular function and show that your function is an isomorphism.]

   (b) Find a new structure that is isomorphic to the structure given in Example 1.6.2. Prove that the structures are isomorphic.

   (c) Find two different structures for a particular language and prove that they are not isomorphic.

(d) Find two different structures for a particular language such that the structures have the same number of elements in their universes but they are still not isomorphic. Prove they are not isomorphic.

6. Take the language of Example 1.6.4 and let C be the set of all $\mathcal{L}$-terms. Create an $\mathcal{L}$-structure $\mathfrak{C}$ by using this universe in such a way that the interpretation of a term t is *not* equal to t.

7. If we take the language $\mathcal{L}_{NT}$, we can create a Henkin structure for that language in the same way as in Example 1.6.4. Do so. Consider the $\mathcal{L}_{NT}$-formula $S0+S0 = SS0$. Is this formula "true" (whatever that means) in your structure? Justify your answer.

1.7 Truth in a Structure

It is at last time to tie together the syntax and the semantics. We have some formal rules about what constitutes a language, and we can identify the terms, formulas, and sentences of a language. We can also identify $\mathcal{L}$-structures for a given language $\mathcal{L}$. In this section we will decide what it means to say that an $\mathcal{L}$-formula ϕ is *true* in an $\mathcal{L}$-structure $\mathfrak{A}$.

To begin the process of tying together the symbols with the structures, we will introduce assignment functions. These assignment functions will formalize what it means to interpret a term or a formula in a structure.

Definition 1.7.1. If $\mathfrak{A}$ is an $\mathcal{L}$-structure, a **variable assignment function into** $\mathfrak{A}$ is a function s that assigns to each variable an element of the universe A. So a variable assignment function into $\mathfrak{A}$ is any function with domain *Vars* and codomain A.

Variable assignment functions need not be injective or bijective. For example, if we work with $\mathcal{L}_{NT}$ and the standard structure $\mathfrak{N}$, then the function s defined by $s(v_i) = i$ is a variable assignment function, as is the function s' defined by

$$s'(v_i) = \text{the smallest prime number that does not divide } i.$$

We will have occasion to want to fix the value of the assignment function s for certain variables.

Definition 1.7.2. If s is a variable assignment function into $\mathfrak{A}$ and x is a variable and $a \in A$, then $s[x|a]$ is the variable assignment function into $\mathfrak{A}$ defined as follows:

$$s[x|a](v) = \begin{cases} s(v) & \text{if } v \neq x \\ a & \text{if } v = x. \end{cases}$$

We call the function $s[x|a]$ an **x-modification of the assignment function** s.

So an x-modification of s is just like s, except that the variable x is assigned to a particular element of the universe.

What we will do next is extend a variable assignment function s to a term assignment function, $\overline{s}$. This function will assign an element of the universe to each term of the language $\mathcal{L}$.

Definition 1.7.3. Suppose that $\mathfrak{A}$ is an $\mathcal{L}$-structure and s is a variable assignment function into $\mathfrak{A}$. The function $\overline{s}$, called the **term assignment function generated by** s, is the function with domain consisting of the set of $\mathcal{L}$-terms and codomain A defined recursively as follows:

1. If t is a variable, $\overline{s}(t) = s(t)$.
2. If t is a constant symbol c, then $\overline{s}(t) = c^{\mathfrak{A}}$.
3. If t is $ft_1t_2\ldots t_n$, then $\overline{s}(t) = f^{\mathfrak{A}}(\overline{s}(t_1), \overline{s}(t_2), \ldots, \overline{s}(t_n))$.

There is a potential problem that needs to be addressed. It is not immediately clear, given a variable assignment function s, that there *is* an extension of s to a term assignment function $\overline{s}$. For example, maybe I can find a term t such that both rules (2) and (3) apply to t, and thus there might be two different possibilities for $\overline{s}(t)$. Fortunately, the unique readability for terms that you proved back in Section 1.3 (Problem 7) provides the assurance that this cannot happen. The formal statement of this fact, the Recursion Theorem, would take us too far afield at this point. The interested reader can find the theorem and its proof in [Enderton 72].

Although we will be primarily interested in truth of sentences, we will first describe truth (or satisfaction) for arbitrary formulas, relative to an assignment function.

Definition 1.7.4. Suppose that $\mathfrak{A}$ is an $\mathcal{L}$-structure, ϕ is an $\mathcal{L}$-formula, and $s : Vars \to A$ is an assignment function. We will say that $\mathfrak{A}$ **satisfies** ϕ **with assignment** s, and write $\mathfrak{A} \models \phi[s]$, in the following circumstances:

1. If ϕ is $= t_1 t_2$ and $\bar{s}(t_1)$ is the same element of the universe A as $\bar{s}(t_2)$, or
2. If ϕ is $R t_1 t_2 \ldots t_n$ and $\langle \bar{s}(t_1), \bar{s}(t_2), \ldots, \bar{s}(t_n) \rangle \in R^{\mathfrak{A}}$, or
3. If ϕ is $(\neg\alpha)$ and $\mathfrak{A} \not\models \alpha[s]$, (where $\not\models$ means "does not satisfy") or
4. If ϕ is $(\alpha \vee \beta)$ and $\mathfrak{A} \models \alpha[s]$, or $\mathfrak{A} \models \beta[s]$ (or both), or
5. If ϕ is $(\forall x)(\alpha)$ and, for each element a of A, $\mathfrak{A} \models \alpha[s(x|a)]$.

If Γ is a set of $\mathcal{L}$-formulas, we say that $\mathfrak{A}$ satisfies Γ with assignment s, and write $\mathfrak{A} \models \Gamma[s]$ if for each $\gamma \in \Gamma$, $\mathfrak{A} \models \gamma[s]$.

Chaff: Notice that the symbol $\models$ is *not* part of the language $\mathcal{L}$. Rather, $\models$ is a metalinguistic symbol that we use to talk about formulas in the language and structures for the language.

Chaff: Also notice that we have at last tied together the syntax and the semantics of our language! The definition above is the place where we formally put the meanings on the symbols that we will use, so that $\vee$ means "or" and $\forall$ means "for all."

Example 1.7.5. Let us work with the empty language, so $\mathcal{L}$ has no constant symbols, no function symbols, and no relation symbols. So an $\mathcal{L}$-structure is simply a nonempty set, and let us consider the $\mathcal{L}$-structure $\mathfrak{A}$, where $A = \{\text{Humphrey}, \text{Ingrid}\}$. Now if we think about the formula $x = y$ and look at the assignment function s, where $s(x)$ is Humphrey and $s(y)$ is also Humphrey. If we ask whether $\mathfrak{A} \models x = y[s]$, we have to check whether $\bar{s}(x)$ is the same element of A as $\bar{s}(y)$. Since the two objects are identical, the formula is true.

To emphasize this, the formula $x = y$ can be true in some universes with some assignment functions. Although the variables x and y are distinct, the truth or falsity of the formula depends *not* on the variables (which are not equal) but rather, on which elements of the structure the variables denote, the *values* of the variables (which are equal for this example). Of course, there are other assignment functions and other structures that make our formula false. I am sure you can think of some.

To talk about the truth or falsity of a *sentence* in a structure, we will take our definition of satisfaction relative to an assignment function and prove that for sentences, the choice of the assignment function is inconsequential. Then we will say that a sentence σ is true in a structure $\mathfrak{A}$ if and only if $\mathfrak{A} \models \sigma[s]$ for any (and therefore all) variable assignment functions s.

> *Chaff:* The next couple of proofs are proofs by induction on the complexity of terms or formulas. You may want to reread the proof of Theorem 1.4.2 on page 18 if you find these difficult.

Lemma 1.7.6. *Suppose that s_1 and s_2 are variable assignment functions into a structure $\mathfrak{A}$ such that $s_1(v) = s_2(v)$ for every variable v in the term t. Then $\overline{s_1}(t) = \overline{s_2}(t)$.*

Proof: We use induction on the complexity of the term t. If t is either a variable or a constant symbol, the result is immediate. If t is $ft_1t_2 \ldots t_n$, then as $\overline{s_1}(t_i) = \overline{s_2}(t_i)$ for $1 \leq i \leq n$ by the inductive hypothesis, the definition of $\overline{s_1}(t)$ and the definition of $\overline{s_2}(t)$ are identical, and thus $\overline{s_1}(t) = \overline{s_2}(t)$. ■

Proposition 1.7.7. *Suppose that s_1 and s_2 are variable assignment functions into a structure $\mathfrak{A}$ such that $s_1(v) = s_2(v)$ for every free variable v in the formula ϕ. Then $\mathfrak{A} \models \phi[s_1]$ if and only if $\mathfrak{A} \models \phi[s_2]$.*

Proof: We use induction on the complexity of ϕ. If ϕ is $= t_1t_2$, then the free variables of ϕ are exactly the variables that occur in ϕ. Thus Lemma 1.7.6 tells us that $\overline{s_1}(t_1) = \overline{s_2}(t_1)$ and $\overline{s_1}(t_2) = \overline{s_2}(t_2)$, meaning that they are the same element of the universe A, so $\mathfrak{A} \models (= t_1t_2)[s_1]$ if and only if $\mathfrak{A} \models (= t_1t_2)[s_2]$, as needed.

The other base case, if ϕ is $Rt_1t_2 \ldots t_n$, is similar and is left as part of Exercise 6.

To begin the first inductive clause, if ϕ is $\neg\alpha$, notice that the free variables of ϕ are exactly the free variables of α, so s_1 and s_2 agree on the free variables of α. By the inductive hypothesis, $\mathfrak{A} \models \alpha[s_1]$

if and only if $\mathfrak{A} \models \alpha[s_2]$, and thus (by the definition of satisfaction), $\mathfrak{A} \models \phi[s_1]$ if and only if $\mathfrak{A} \models \phi[s_2]$. The second inductive clause, if ϕ is $\alpha \lor \beta$, is another part of Exercise 6.

If ϕ is $(\forall x)(\alpha)$, we first note that the only variable that might be free in α that is not free in ϕ is x. Thus, if $a \in A$, the assignment functions $s_1[x|a]$ and $s_2[x|a]$ agree on all of the free variables of α. Therefore, by inductive hypothesis, for each $a \in A$, $\mathfrak{A} \models \alpha[s_1[x|a]]$ if and only if $\mathfrak{A} \models \alpha[s_2[x|a]]$. So, by Definition 1.7.4, $\mathfrak{A} \models \phi[s_1]$ if and only if $\mathfrak{A} \models \phi[s_2]$. This finishes the last inductive clause, and our proof. ■

Corollary 1.7.8. *If σ is a sentence in the language $\mathcal{L}$ and $\mathfrak{A}$ is an $\mathcal{L}$-structure, either $\mathfrak{A} \models \sigma[s]$ for all assignment functions s, or $\mathfrak{A} \models \sigma[s]$ for no assignment function s.*

Proof: There *are* no free variables in σ, so if s_1 and s_2 are two assignment functions, they agree on all of the free variables of σ, there just aren't all that many of them. So by Proposition 1.7.7, $\mathfrak{A} \models \sigma[s_1]$ if and only if $\mathfrak{A} \models \sigma[s_2]$, as needed. ■

Definition 1.7.9. If ϕ is a formula in the language $\mathcal{L}$ and $\mathfrak{A}$ is an $\mathcal{L}$-structure, we say that $\mathfrak{A}$ is a **model** of ϕ, and write $\mathfrak{A} \models \phi$, if and only if $\mathfrak{A} \models \phi[s]$ for every assignment function s. If Φ is a set of $\mathcal{L}$-formulas, we will say that $\mathfrak{A}$ models Φ, and write $\mathfrak{A} \models \Phi$, if and only if $\mathfrak{A} \models \phi$ for each $\phi \in \Phi$.

Notice that if σ is a *sentence*, then $\mathfrak{A} \models \sigma$ if and only if $\mathfrak{A} \models \sigma[s]$ for *any* assignment function s. In this case we will say that the sentence σ is **true in $\mathfrak{A}$**.

Example 1.7.10. Let's work in $\mathcal{L}_{NT}$, and let

$$\mathfrak{N} = \langle \mathbb{N}, 0, S, +, \cdot, E, < \rangle$$

be the standard structure. Let s be the variable assignment function that assigns v_i to the number $2i$. Now let the formula $\phi(v_1)$ be $v_1 + v_1 = SSSS0$.

To show that $\mathfrak{N} \models \phi[s]$, notice that

$$\begin{aligned} \bar{s}(v_1 + v_1) \quad & \text{is} \quad +^{\mathfrak{N}}(\bar{s}(v_1), \bar{s}(v_1)) \\ & \text{is} \quad +^{\mathfrak{N}}(2, 2) \\ & \text{is} \quad 4 \end{aligned}$$

while

$$\bar{s}(SSSS0) \quad \text{is} \quad S^{\mathfrak{N}}(S^{\mathfrak{N}}(S^{\mathfrak{N}}(S^{\mathfrak{N}}(0^{\mathfrak{N}}))))$$
$$\text{is} \quad 4.$$

Now, in the same setting, consider σ, the sentence

$$(\forall v_1)\neg(\forall v_2)\neg(v_1 = v_2 + v_2),$$

which states that everything is even. [That is hard to see unless you know to look for that $\neg(\forall v_2)\neg$ and to read it as $(\exists v_2)$. See the last couple of paragraphs of this section.] You know that σ is false in the standard structure, but to show how the formal argument goes, let s be any variable assignment function and notice that

$$\begin{aligned}
\mathfrak{N} \models \sigma[s] \quad \text{iff} \quad & \text{For every } a \in \mathbb{N},\ \mathfrak{N} \models \neg(\forall v_2)\neg(v_1 = v_2 + v_2)s[v_1|a] \\
\text{iff} \quad & \text{For every } a \in \mathbb{N},\ \mathfrak{N} \not\models (\forall v_2)\neg(v_1 = v_2 + v_2)s[v_1|a] \\
\text{iff} \quad & \text{For every } a \in \mathbb{N}, \text{ there is a } b \in \mathbb{N}, \\
& \mathfrak{N} \models v_1 = v_2 + v_2 s[v_1|a][v_2|b].
\end{aligned}$$

Now, if we consider the case when a is the number 3, it is perfectly clear that there is no such b, so we have shown $\mathfrak{N} \not\models \sigma[s]$. Then, by Definition 1.7.9, we see that the sentence σ is false in the standard structure. As you well knew.

When you were introduced to symbolic logic, you were probably told that there were five connectives. In the mathematics that you have learned recently, you have been using two quantifiers. I hope you have noticed that we have not used all of those symbols in this book, but it is now time to make those symbols available. Rather than adding the symbols to our language, however, we will introduce them as abbreviations. This will help us to keep our proofs slightly less complex (as our inductive proofs will have fewer cases) but will still allow us to use the more familiar symbols, at least as shorthand.

Thus, let us agree to use the following abbreviations in constructing $\mathcal{L}$-formulas: We will write $(\alpha \wedge \beta)$ instead of $(\neg((\neg\alpha) \vee (\neg\beta)))$, $(\alpha \rightarrow \beta)$ instead of $((\neg\alpha) \vee \beta)$, and $(\alpha \leftrightarrow \beta)$ instead of $((\alpha \rightarrow \beta) \wedge (\beta \rightarrow \alpha))$. We will also introduce our missing existential quantifier as an abbreviation, writing $(\exists x)(\alpha)$ instead of $(\neg(\forall x)(\neg\alpha))$. It is an easy exercise to check that the introduced connectives $\wedge$, $\rightarrow$, and $\leftrightarrow$ behave as you would expect them to. Thus $\mathfrak{A} \models (\alpha \land \beta)[s]$ if and only if both $\mathfrak{A} \models \alpha[s]$ and $\mathfrak{A} \models \beta[s]$. The existential quantifier is only slightly more difficult. See Exercise 7.

1.7.1 Exercises

1. I suggested after Definition 1.5.3 that the truth or falsity of the sentences $1+1=2$ and $(\forall x)(x+1=x)$ might not be automatic. Find a structure for the language discussed there that makes the sentence $1+1=2$ true. Find another structure where $1+1=2$ is false. Prove your assertions. Then show that you can find a structure where $(\forall x)(x+1=x)$ is true, and another structure where it is false.

2. Let the language $\mathcal{L}$ be $\{S, <\}$, where S is a unary function symbol and $<$ is a binary relation symbol. Let ϕ be the formula $(\forall x)(\exists y)(Sx < y)$.

   (a) Find an $\mathcal{L}$-structure $\mathfrak{A}$ such that $\mathfrak{A} \models \phi$.

   (b) Find an $\mathcal{L}$-structure $\mathfrak{B}$ such that $\mathfrak{B} \models (\neg\phi)$.

   (c) Prove that your answer to part (a) or part (b) is correct.

   (d) Write an $\mathcal{L}$-sentence that is true in a structure $\mathfrak{A}$ if and only if the universe A of $\mathfrak{A}$ consists of exactly two elements.

3. Consider the language and structure of Example 1.6.4. Write two nontrivial sentences in the language, one of which is true in the structure and one of which (not the denial of the first) is false in the structure. Justify your assertions.

4. Consider the sentence σ: $(\forall x)(\exists y)\left[x < y \rightarrow x+1=y\right]$. Find two structures for a suitable language, one of which makes σ true, and the other of which makes σ false.

5. One more bit of shorthand. Assume that the language $\mathcal{L}$ contains the binary relation symbol $\in$, which you are intending to use to mean the elementhood relation (so $p \in q$ will mean that p is an element of q). Often, it is the case that you want to claim that $\phi(x)$ is true for every element of a set b. Of course, to do this you could write

$$(\forall x)\left[(x \in b) \rightarrow \phi(x)\right].$$

We will abbreviate this formula as

$$(\forall x \in b)(\phi(x)).$$

Similarly, $(\exists x \in b)(\phi(x))$ will be an abbreviation for the formula $(\exists x)\big[(x \in b) \land \phi(x)\big]$. Notice that this formula has a conjunction where the previous formula had an implication! We do that just to see if you are paying attention. (Well, if you think about what the abbreviations are supposed to mean, you'll see that the change is necessary. We'll have to do something else just to see if you're paying attention.)

Now suppose that $\mathfrak{A}$ is a structure for the language of set theory. So $\mathcal{L}$ has only this one binary relation symbol, $\in$, which is interpreted as the elementhood relation. Suppose, in addition, that $A = \{u, v, w, \{u\}, \{u, v\}, \{u, v, w\}\}$. In particular, notice that there is no element x of A such that $x \in x$. Consider the sentence

$$(\forall y \in y)(\exists x \in x)(x = y).$$

Is this sentence true or false in $\mathfrak{A}$?

6. Fill in the details to complete the proof of Proposition 1.7.7.

7. Show that $\mathfrak{A} \models (\exists x)(\alpha)[s]$ if and only if there is an element $a \in A$ such that $\mathfrak{A} \models \alpha[s[x|a]]$.

1.8 Substitutions and Substitutability

Suppose you knew that the sentence $\forall x \phi(x)$ was true in a particular structure $\mathfrak{A}$. Then, if c is a constant symbol in the language, you would certainly expect $\phi(c)$ to be true in $\mathfrak{A}$ as well. What we have done is substitute the constant symbol c for the variable x. This seems perfectly reasonable, although there are times when you do have to be careful.

Suppose that $\mathfrak{A} \models \forall x \exists y \neg(x = y)$. This sentence is, in fact, true in any structure $\mathfrak{A}$ such that A has at least two elements. If we then proceed to replace the variable x by the variable u, we get the statement $\exists y \neg(u = y)$, which will still be true in $\mathfrak{A}$, no matter what value we give to the variable u. If, however, we take our original formula and replace x by y, then we find ourselves looking at $\exists y \neg(y = y)$, which will be false in any structure. So by a poor choice of substituting variable, we have changed the truth value of our formula. The rules of substitutability that we will discuss in this section are designed to help us avoid this problem, the problem of attempting to substitute a term inside a quantifier that binds a variable involved in the term.

We begin by defining exactly what we mean when we substitute a term t for a variable x in either a term u or a formula ϕ.

Definition 1.8.1. Suppose that u is a term, x is a variable, and t is a term. We define the term u_t^x (read "u with x replaced by t") as follows:

1. If u is a variable not equal to x, then u_t^x is u.
2. If u is x, then u_t^x is t.
3. If u is a constant symbol, then u_t^x is u.
4. If u is $fu_1u_2\ldots u_n$, where f is an n-ary function symbol and the u_i are terms, then

$$u_t^x \text{ is } f(u_1)_t^x(u_2)_t^x\ldots(u_n)_t^x.$$

Chaff: In the fourth clause of the definition above and in the first two clauses of the next definition, the parentheses are not really there. However, I personally believe that no one can look at $u_{1t}^{\,x}$ and figure out what it is supposed to mean. So the parentheses have been added in the interest of readability.

For example, if we let t be $g(c)$ and we let u be $f(x,y)+h(z,x,g(x))$, then u_t^x is

$$f(g(c),y)+h(z,g(c),g(g(c))).$$

The definition of substitution into a formula is also by recursion:

Definition 1.8.2. Suppose that ϕ is an $\mathcal{L}$-formula, t is a term, and x is a variable. We define the formula ϕ_t^x (read "ϕ with x replaced by t") as follows:

1. If ϕ is $= u_1u_2$, then ϕ_t^x is $= (u_1)_t^x(u_2)_t^x$.

2. If ϕ is $Ru_1u_2\ldots u_n$, then ϕ^x_t is $R(u_1)^x_t(u_2)^x_t\ldots(u_n)^x_t$.
3. If ϕ is $\neg(\alpha)$, then ϕ^x_t is $\neg(\alpha^x_t)$.
4. If ϕ is $(\alpha \vee \beta)$, then ϕ^x_t is $(\alpha^x_t \vee \beta^x_t)$.
5. If ϕ is $(\forall y)(\alpha)$, then

$$\phi^x_t = \begin{cases} \phi & \text{if } x \text{ is } y \\ (\forall y)(\alpha^x_t) & \text{otherwise.} \end{cases}$$

As an example, suppose that ϕ is the formula

$$P(x,y) \rightarrow \big[(\forall x)(Q(g(x),z)) \vee (\forall y)(R(x,h(x))\big].$$

Then, if t is the term $g(c)$, we get

$$\phi^x_t \text{ is } P(g(c),y) \rightarrow \big[(\forall x)(Q(g(x),z)) \vee (\forall y)(R(g(c),g(g(c)))\big].$$

Having defined what we mean when we substitute a term for a variable, we will now define what it means for a term to be substitutable for a variable in a formula. The idea is that if t is substitutable for x in ϕ, we will not run into the problems discussed at the beginning of this section—we will not substitute a term in such a way that a variable contained in that term is inadvertently bound by a quantifier.

Definition 1.8.3. Suppose that ϕ is an $\mathcal{L}$-formula, t is a term, and x is a variable. We say that t **is substitutable for** x **in** ϕ if

1. ϕ is atomic, or
2. ϕ is $\neg(\alpha)$ and t is substitutable for x in α, or
3. ϕ is $(\alpha \vee \beta)$ and t is substitutable for x in both α and β, or
4. ϕ is $(\forall y)(\alpha)$ and either
   (a) x is not free in ϕ, or
   (b) y does not occur in t and t is substitutable for x in α.

Notice that ϕ^x_t is defined whether or not t is substitutable for x in ϕ. Usually, we will not want to do a substitution unless we check for substitutability, but we have the ability to substitute whether or not it is a good idea. In the next chapter, however, you will often see that certain operations are allowed only if t is substitutable for x in ϕ. That restriction is there for good reason, as we will be concerned with preserving the truth of formulas after performing substitutions.

1.8.1 Exercises

1. For each of the following, write out u_t^x:

(a) u is $\cos x$, t is $\sin y$.

(b) u is y, t is Sy.

(c) u is $\natural(x, y, z)$, t is $423 - w$.

2. For each of the following, first write out ϕ_t^x, then decide if t is substitutable for x in ϕ, and then (if you haven't already) use the definition of substitutability to justify your conclusions.

(a) ϕ is $\forall x(x = y \rightarrow Sx = Sy)$, t is $S0$.

(b) ϕ is $\forall y(x = y \rightarrow Sx = Sy)$, t is Sy.

(c) ϕ is $x = y \rightarrow (\forall x)(Sx = Sy)$, t is Sy.

3. Show that if t is variable-free, then t is always substitutable for x in ϕ.

4. Show that x is always substitutable for x in ϕ.

5. Prove that if x is not free in ψ, then ψ_t^x is ψ.

6. You might think that $(\phi_y^x)_x^y$ is ϕ, but a moment's thought will give you an example to show that this doesn't always work. (What if y is free in ϕ?) Find an example that shows that even if y is not free in ϕ, we can still have $(\phi_y^x)_x^y$ different from ϕ. Under what conditions do we know that $(\phi_y^x)_x^y$ is ϕ?

7. Write a computer program (in your favorite language, or in pseudo-code) that accepts as input a formula ϕ, a variable x, and a term t and outputs "yes" or "no" depending on whether or not t is substitutable for x in ϕ.

1.9 Logical Implication

At first glance it seems that a large portion of mathematics can be broken down into answering questions of the form: If I know this statement is true, is it necessarily the case that this other statement is true? In this section we will formalize that question.

Definition 1.9.1. Suppose that Δ and Γ are sets of $\mathcal{L}$-formulas. We will say that Δ **logically implies** Γ and write $\Delta \models \Gamma$ if for every $\mathcal{L}$-structure $\mathfrak{A}$, if $\mathfrak{A} \models \Delta$, then $\mathfrak{A} \models \Gamma$.

This definition is a little bit tricky. It says that if Δ is true in $\mathfrak{A}$, then Γ is true in $\mathfrak{A}$. Remember, for Δ to be true in $\mathfrak{A}$, it must be the case that $\mathfrak{A} \models \Delta[s]$ for *every* assignment function s. See Exercise 4.

If $\Gamma = \{\gamma\}$ is a set consisting of a single formula, we will write $\Delta \models \gamma$ rather than the official $\Delta \models \{\gamma\}$.

Definition 1.9.2. An $\mathcal{L}$-formula ϕ is said to be **valid** if $\emptyset \models \phi$, in other words, if ϕ is true in every $\mathcal{L}$-structure with every assignment function s. In this case, we will write $\models \phi$.

> *Chaff:* It doesn't seem like it would be easy to check whether $\Delta \models \Gamma$. To do so directly would mean that we would have to examine every possible $\mathcal{L}$-structure and every possible assignment function s, of which there will be many.
>
> I'm also sure that you've noticed that this double turnstyle symbol, $\models$, is getting a lot of use. Just remember that if there is a structure on the left, $\mathfrak{A} \models \sigma$, we are discussing truth in a single structure. If there is a set of sentences on the left, $\Gamma \models \sigma$, then we are discussing logical implication.

Example 1.9.3. Let $\mathcal{L}$ be the language consisting of a single binary relation symbol, P, and let σ be the sentence $(\exists y \forall x P(x,y)) \to (\forall x \exists y P(x,y))$. We show that σ is valid.

So let $\mathfrak{A}$ be any $\mathcal{L}$-structure and let $s : Vars \to A$ be any assignment function. We must show that

$$\mathfrak{A} \models \Big[(\exists y \forall x P(x,y)) \to (\forall x \exists y P(x,y))\Big][s].$$

Assume that $\mathfrak{A} \models (\exists y \forall x P(x,y))\,[s]$. (If $\mathfrak{A}$ does not model this sentence, then we know by the definition of $\to$ that $\mathfrak{A} \models \sigma[s]$.)

Since we know that $\mathfrak{A} \models (\exists y \forall x P(x,y))\,[s]$, we know that there is an element of the universe, a, such that $\mathfrak{A} \models \forall x P(x,y)[s[y|a]]$. And so, again by the definition of satisfaction, we know that if b is any element of A, $\mathfrak{A} \models P(x,y)\,[(s[y|a])\,[x|b]]$. If we chase through the definition of satisfaction (Definition 1.7.4) and of the various assignment functions, this means that for our one fixed a, the ordered pair $\langle b, a\rangle \in P^{\mathfrak{A}}$ for any choice of $b \in A$, .

We have to prove that $\mathfrak{A} \models (\forall x \exists y P(x,y))\,[s]$. As the statement of interest is universal, we must show that, if c is an arbitrary element of A, $\mathfrak{A} \models \exists y P(x,y)[s[x|c]]$, which means that we must produce an element of the universe, d, such that $\mathfrak{A} \models P(x,y)\,[(s[x|c])\,[y|d]]$. Again, from the definition of satisfaction this means that we must find a $d \in A$ such that $\langle c, d\rangle \in P^{\mathfrak{A}}$. Fortunately, we have such a d in hand, namely a. As we know $\langle c, a\rangle \in P^{\mathfrak{A}}$, we have shown $\mathfrak{A} \models (\forall x \exists y P(x,y))\,[s]$, and we are finished.

1.9.1 Exercises

1. Show that $\{\alpha, \alpha \to \beta\} \models \beta$ for any formulas α and β. Translate this result into everyday English. Or Swedish, if you prefer.

2. Show that the formula $x = x$ is valid. Show that the formula $x = y$ is not valid. What can you prove about the formula $\neg x = y$ in terms of validity?

3. Suppose that ϕ is an $\mathcal{L}$-formula and x is a variable. Prove that ϕ is valid if and only if $(\forall x)(\phi)$ is valid. Thus, if ϕ has free variables x, y, and z, ϕ will be valid if and only if $\forall x \forall y \forall z \phi$ is valid. The sentence $\forall x \forall y \forall z \phi$ is called the **universal closure** of ϕ.

4. (a) Assume that $\models (\phi \to \psi)$. Show that $\phi \models \psi$.

(b) Suppose that ϕ is $x < y$ and ψ is $z < w$. Show that $\phi \models \psi$ but $\not\models (\phi \to \psi)$. (The slash through $\models$ means "does not logically imply.")

[This exercise shows that the two possible ways to define logical equivalence are not equivalent. The strong form of the definitions says that ϕ and ψ are logically equivalent if $\models (\phi \to \psi)$ and $\models (\psi \to \phi)$. The weak form of the definition states that ϕ and ψ are logically equivalent if $\phi \models \psi$ and $\psi \models \phi$.]

1.10 Summing Up, Looking Ahead

What we have tried to do in this first chapter is to introduce the concepts of formal languages and formal structures. I hope that you will agree that you have seen many mathematical structures in the past, even though you may not have called them structures at the time. By formalizing what we mean when we say that a formula is true in a structure, we will be able to tie together truth and provability in the next couple of chapters.

You might be at a point where you are about to throw your hands up in disgust and say, "Why does any of this matter? I've been doing mathematics for over ten years without worrying about structures or assignment functions, and I have been able to solve problems and succeed as a mathematician so far." Allow me to assure you that the effort and the almost unreasonable precision that we are imposing on our exposition will have a payoff in later chapters. The major theorems that we wish to prove are theorems about the existence or nonexistence of certain objects. To prove that you cannot express a certain idea in a certain language, we have to *know*, with an amazing amount of exactitude, what a language is and what structures are. Our goals are some theorems that are easy to state incorrectly, so by being precise about what we are saying, we will be able to make (and prove) claims that are truly revolutionary.

Since we will be talking about the existence and nonexistence of proofs, we now must turn our attention to defining (yes, precisely) what sorts of things qualify as proofs. That is the topic of the next chapter.

Chapter 2

Deductions

2.1 Naïvely

What is it that makes mathematics different from other academic subjects? What is it that distinguishes a mathematician from a poet, a linguist, a biologist, or a civil engineer? I am sure that you have many answers to that question, not all of which are complimentary to the author of this work or to the mathematics instructors that you have known!

I suggest that one of the things that sets mathematics apart is the insistence upon proof. Mathematical statements are not accepted as true until they have been verified, and verified in a very particular manner. This process of verification is central to the subject and serves to define our field of study in the minds of many. Allow me to quote a famous story from John Aubrey's *Brief Lives*:

> [Thomas Hobbes] was 40 years old before he looked on Geometry; which happened accidentally. Being in a Gentleman's Library, Euclid's Elements lay open and 'twas the 47 El. libri 1 [the Pythagorean Theorem]. He read the Proposition. By G—, sayd he (he would now and then sweare an emphaticall Oath by way of emphasis) this is impossible! So he reads the Demonstration of it, which referred him back to such a Proposition; which proposition he read. That referred him back to another, which he also read. Et sic deinceps [and so on] that at last he was demonstratively convinced of that trueth. This made him in love with Geometry.

Doesn't this match pretty well with your image of a mathematical proof? To prove a proposition, you start from some first principles, derive some results from those axioms, then, using those axioms and results, push on to prove other results. This is a technique that you have seen in geometry courses, college mathematics courses, and in the first chapter of this book.

Our goal in this chapter will be to define, precisely, something called a deduction. You probably haven't seen a deduction before, and you aren't going to see very many of them after this chapter is over, but our idea will be that any mathematical proof should be able to be translated into a (probably very long) deduction. This will be crucial in our interpretation of the results of Chapters 3 and 5, where we will discuss the existence and nonexistence of certain deductions, and interpret those results as making claims about the existence and nonexistence of mathematical proofs.

If you think about what a proof is, you probably will come up with a characterization along the lines of: A proof is a sequence of statements, each one of which can be justified by referring to previous statements. This is a perfectly reasonable starting point, and it brings us to the main difficulty we will have to address as we move from an informal understanding of what constitutes a proof to a formal definition of a deduction: What do you mean by the word *justified*?

Our answer to this question will come in three parts. We will start by specifying a set Λ of $\mathcal{L}$-formulas, which will be called the logical axioms. Logical axioms will be deemed to be "justified" in any deduction. Depending on the situation at hand, we will then specify a set of nonlogical axioms, Σ. Finally, we will develop some rules of inference, which will be pairs $\langle \Gamma, \phi \rangle$, where Γ is a finite set of formulas and ϕ is a formula. Then, if α is a formula, we will say that a deduction of α from Σ is a finite list of formulas $\phi_1, \phi_2, \ldots, \phi_n$ such that ϕ_n is α and for each i, ϕ_i is justified by virtue of being either a logical axiom ($\phi_i \in \Lambda$), a nonlogical axiom ($\phi_i \in \Sigma$), or the conclusion of one of our rules of inference, $\langle \Gamma, \phi_i \rangle$, where $\Gamma \subseteq \{\phi_1, \phi_2, \ldots, \phi_{i-1}\}$.

The proofs that you have seen in your mathematical career have had a couple of nice properties. The first of these is that proofs are easy to follow. (OK, they aren't *always* easy to follow, but they are supposed to be.) This doesn't mean that it is easy to *discover* a proof, but rather that if someone is showing you a proof, it should be easy to follow the steps of the proof and to understand why the proof is correct. The second admirable property of proofs is that when you prove something, you know that it is true! Our definition of deduction will be designed to make sure that deductions, too, will be easily checkable and will preserve truth.

In order to do this, we will impose the following restrictions on our logical axioms and rules of inference:

1. There will be an algorithm (i.e., a mechanical procedure) that will decide, given a formula θ, whether or not θ is a logical axiom.

2. There will be an algorithm that will decide, given a finite set of formulas Γ and a formula θ, whether or not $\langle \Gamma, \theta \rangle$ is a rule of inference.

3. For each rule of inference $\langle \Gamma, \theta \rangle$, Γ will be a finite set of formulas.

4. Each logical axiom will be valid.

5. Our rules of inference will preserve truth. In other words, for each rule of inference $\langle \Gamma, \theta \rangle$, $\Gamma \models \theta$.

The idea here is that although it may require no end of brilliance and insight to discover a deduction of a formula α, there should be *no* brilliance and *no* insight required to check whether an alleged deduction of α is, in fact, a deduction of α. To check whether a deduction is correct will be such a simple procedure that it could be programmed into a computer. Furthermore, we will be certain that if a deduction of α from Σ is given, and if we look at a mathematical structure $\mathfrak{A}$ such that $\mathfrak{A} \models \Sigma$, then we will be certain that $\mathfrak{A} \models \alpha$. This is what we mean when we say that our deductions will preserve truth.

2.2 Deductions

We begin by fixing a language $\mathcal{L}$. Also assume that we have been given a fixed set of $\mathcal{L}$-formulas, Λ, called the set of logical axioms, and a set of ordered pairs $\langle \Gamma, \phi \rangle$, called the rules of inference. (We will specify which formulas are elements of Λ and which ordered pairs are rules of inference in the next two sections.) A deduction is going to be a finite sequence, or list, of $\mathcal{L}$-formulas with certain properties.

Definition 2.2.1. Suppose that Σ is a collection of $\mathcal{L}$-formulas and D is a finite sequence $\langle \phi_1, \phi_2, \ldots, \phi_n \rangle$ of $\mathcal{L}$-formulas. We will say that D is a **deduction from** Σ if for each i, $1 \leq i \leq n$, either

1. $\phi_i \in \Lambda$ (ϕ_i is a logical axiom), or

2. $\phi_i \in \Sigma$ (ϕ_i is a nonlogical axiom), or

3. There is a rule of inference $\langle \Gamma, \phi_i \rangle$ such that $\Gamma \subseteq \{\phi_1, \phi_2, \ldots, \phi_{i-1}\}$.

If there is a deduction from Σ, the last line of which is the formula ϕ, we will call this a **deduction from** Σ **of** ϕ, and write $\Sigma \vdash \phi$.

> *Chaff:* Well, we have now established what we mean by the word *justified*. In a deduction we are allowed to write down any $\mathcal{L}$-formula that we like, as long as that formula is either a logical axiom or is listed explicitly in a collection Σ of nonlogical axioms. Any formula that we write in a deduction that is *not* an axiom must arise from previous formulas in the deduction via a rule of inference.
>
> You may have gathered that there are many different deductive systems, depending on the choices that are made for Λ, and the rules of inference. As a general rule, a deductive system will either have lots of rules of inference and few logical axioms, or not too many rules and a lot of axioms. In developing the deductive system for us to use in this book, we attempt to pursue a middle course.
>
> Also notice that $\vdash$ is another metalinguistic symbol. It is not part of the language $\mathcal{L}$.

Example 2.2.2. Suppose, for starters, that we don't want to make any assumptions. So, let $\Sigma = \emptyset$, let $\Lambda = \emptyset$, and write down a deduction from Σ. Don't be shy. Go ahead. I'll wait.

Still nothing? Right. There are no deductions from the empty set of axioms. (Actually, after we set up our rules of inference, there will be some deductions from the empty set of axioms, but that comes later.) This is a problem that the English logician Bertrand Russell found particularly annoying as he began to learn mathematics.

> At the age of eleven, I began Euclid, with my brother as my tutor. This was one of the great events of my life, as dazzling as first love. I had not imagined that there was anything so delicious in the world. ... From that moment until Whitehead and I finished *Principia Mathematica*, when I was thirty-eight, mathematics was my chief interest, and my chief source of happiness. Like all happiness, however, it was not unalloyed. I had been told that Euclid proved things, and was much disappointed that he started with axioms. At first I refused to accept them unless my brother could offer me some reason for doing so, but he said: "If you don't accept them we cannot go on," and as I wished to go on, I reluctantly admitted them *pro tem*. The doubt as to the premisses of mathematics which I felt at that moment remained with me, and determined the course of my subsequent work. [Russell 67, p. 36]

What we have managed to do with our definition of deduction, though, is to be up front about our need to make assumptions, and we will acknowledge our axiom set in every deduction that we write.

Example 2.2.3. Let us work in the language $\mathcal{L} = \{P\}$, where P is a binary relation symbol. Let Σ, our set of axioms, be

$$\begin{aligned}\Sigma = \{&\forall x P(x, x),\\ &P(u, v),\\ &P(u, v) \rightarrow P(v, u),\\ &P(v, u) \rightarrow P(u, u)\}.\end{aligned}$$

We will let $\Lambda = \emptyset$ for now. We also need to have a set of rules of inference. So temporarily let our set of rules of inference be

$$\{\langle\{\alpha, \alpha \rightarrow \beta\}, \beta\rangle \mid \alpha \text{ and } \beta \text{ are formulas of } \mathcal{L}\}.$$

This is just the rule modus ponens, which says that from the formulas α and $\alpha \rightarrow \beta$ we may conclude β.

Now we can write a deduction from Σ of the formula $P(u, u)$, as follows:

$$
\begin{aligned}
&P(u, v) \\
&P(u, v) \to P(v, u) \\
&P(v, u) \\
&P(v, u) \to P(u, u) \\
&P(u, u).
\end{aligned}
$$

You can easily see that every formula in our deduction is either explicitly listed among the elements of our axiom set Σ, or follows from modus ponens from previously listed formulas in the deduction.

Notice, however, that we cannot use the universal statement $\forall x P(x, x)$ to derive our needed formula $P(u, u)$. Even a statement that seems like it ought to follow from our axioms, $P(v, v)$, for example, will not be deducible from Σ until we either add to our rules of inference or include some additional axioms. Our definition of a deduction is very limiting—we cannot even use standard logical tricks such as universal instantiation [from $\forall x \, blah(x)$ deduce $blah(t)$]. These logical axioms will be gathered together in Section 2.3.

Chaff: It is really tempting here to write down the incorrect deduction

$$
\begin{aligned}
&\forall x P(x, x) \\
&P(u, u).
\end{aligned}
$$

Please don't say things like that until we have built our collection of logical axioms. Remember, what we are trying to do here is to have a definition of deduction that is entirely syntactic, that does not depend on the meanings of the symbols. Where you are likely to run into trouble is when you start thinking too much about the meanings of the things that you write down. Our definition gives us deductions that are easily verifiable: Given an alleged deduction from Σ, as long as we can decide what formulas are in Σ, we can decide if the alleged deduction is correct. In fact, we could easily program a computer to check the deduction for us. However, this ease in verification comes with a price: Deductions are difficult to write and hard to motivate.

Definition 2.2.1 is a "bottom-up" definition. It defines a deduction in terms of its parts. Another way to define a collection of things is to take a "top-down" approach. The next proposition does just that, by showing that we can think of the collection of deductions from Σ (called Thm_Σ) as the closure of the collection of axioms under the application of the rules of inference.

Proposition 2.2.4. *Fix sets of $\mathcal{L}$-formulas Σ and Λ and a collection of rules of inference. The set $Thm_\Sigma = \{\phi \mid \Sigma \vdash \phi\}$ is the smallest set C such that*

1. $\Sigma \subseteq C$.

2. $\Lambda \subseteq C$.

3. *If $\langle \Gamma, \theta \rangle$ is a rule of inference and $\Gamma \subseteq C$, then $\theta \in C$.*

Proof: This proposition makes two separate claims about the set Thm_Σ. The first claim is that Thm_Σ satisfies the three criteria. The second claim is that Thm_Σ is the *smallest* set to satisfy the criteria. We tackle these claims one at a time.

First, let us look at the criteria in order, and make sure that Thm_Σ satisfies them. So to begin, we must show that $\Sigma \subseteq \text{Thm}_\Sigma$. But certainly if $\sigma \in \Sigma$, there is a deduction-from-Σ of σ, for example this one-line deduction: σ. Similarly, to show that $\Lambda \subseteq \text{Thm}_\Sigma$, we notice that there is a one-line deduction of any $\lambda \in \Lambda$. To finish this part of the proof, we must show that if $\langle \Gamma, \theta \rangle$ is a rule of inference and $\Gamma \subseteq \text{Thm}_\Sigma$, then $\theta \in \text{Thm}_\Sigma$. But to produce a deduction-from-Σ of θ, all we have to do is write down deductions of each of the γ's in Γ, followed by the formula θ. This is a valid deduction, as θ follows from Γ by the rule of inference $\langle \Gamma, \theta \rangle$. Thus Thm_Σ satisfies the three criteria of the proposition.

Now we must show that Thm_Σ is the smallest such set. This is quite easy to prove once you figure out what you have to do. What is claimed is that if C is a collection of formulas satisfying the given requirements, then $\text{Thm}_\Sigma \subseteq C$. So we assume that C is a class satisfying the conditions, and we attempt to show that every element of Thm_Σ is in C.

If $\phi \in \text{Thm}_\Sigma$, there is a deduction from Σ with last line ϕ. If the entry ϕ is justified by virtue of ϕ being either a logical or nonlogical axiom, then ϕ is explicitly included in the set C. If ϕ is justified by reference to a rule of inference $\langle \Gamma, \phi \rangle$, then each $\gamma \in \Gamma$ is an element of C (this is really a proof by induction, and here is where we use the inductive hypothesis), and thus, by the third requirement on C, $\phi \in C$, as needed.

Since $\text{Thm}_\Sigma \subseteq C$ for all such sets C, Thm_Σ is the *smallest* such set, as claimed. ■

Here is what we will do in the next few sections: We will define Λ, the fixed set of logical axioms; we will establish our collection of rules of inference; we will prove some results about deductions; and finally, we will discuss some examples of sets of nonlogical axioms.

2.2.1 Exercises

1. Let the collection of nonlogical axioms be

$$\Sigma = \{(A(x) \land A(x)) \to B(x,y), A(x), B(x,y) \to A(x)\},$$

and let the rule of inference be modus ponens, as in Example 2.2.3. For each of the following, decide if it is a deduction. If it is not a deduction, explain how you know that it is not a deduction.

(a)
$$A(x)$$
$$A(x) \land A(x)$$
$$(A(x) \land A(x)) \to B(x,y)$$
$$B(x,y)$$

(b)
$$B(x,y) \to A(x)$$
$$A(x)$$
$$B(x,y)$$

(c)
$$(A(x) \land A(x)) \to B(x,y)$$
$$B(x,y) \to A(x)$$
$$(A(x) \land A(x)) \to A(x)$$

2. Consider the axiom system Σ of Example 2.2.3. It is implied in that example that there is no deduction from Σ of the formula $P(v,v)$. Prove this fact.

3. Carefully write out the proof of Proposition 2.2.4, worrying about the inductive step. [*Suggestion:* You may want to proceed by induction on the length of the shortest deduction of ϕ.]

4. Let $\mathcal{L}$ be a language that consists of a single unary predicate symbol R, and let B be the infinite set of axioms

$$\begin{aligned} B = \{ & R(x_1), \\ & R(x_1) \rightarrow R(x_2), \\ & R(x_2) \rightarrow R(x_3), \\ & \vdots \\ & R(x_i) \rightarrow R(x_{i+1}), \\ & \vdots \\ & \}. \end{aligned}$$

Using modus ponens as the only rule of inference, prove by induction that $B \vdash R(x_j)$ for each natural number $j \geq 1$.

2.3 The Logical Axioms

Let a first-order language $\mathcal{L}$ be given. In this section we will gather together a collection Λ of logical axioms for $\mathcal{L}$. This set of axioms, though infinite, will be decidable. Roughly this means that if we are given a formula ϕ that is alleged to be an element of Λ, we will be able to decide whether $\phi \in \Lambda$ or $\phi \notin \Lambda$. Furthermore, we could, in principle, design a computer program that would be able to decide membership in Λ in a finite amount of time.

After we have established the set of logical axioms Λ and we want to start doing mathematics, we will want to add additional axioms that are designed to allow us to deduce statements about whatever mathematical system we may have in mind. These will constitute the collection of nonlogical axioms, Σ. For example, if we are working in number theory, using the language $\mathcal{L}_{NT}$, along with the logical axioms Λ we will also want to use other axioms that concern the properties of addition and the ordering relation denoted by the symbol $<$. These additional axioms are the formulas that we will place in Σ. Then, from this expanded set of axioms $\Lambda \cup \Sigma$ we will attempt to write deductions of formulas that make statements of number-theoretic interest. To reiterate: Λ, the set of logical axioms, will be fixed, as will the collection of rules of inference. But the set of nonlogical axioms must be specified for each deduction. In the current section we set out the logical axioms only, dealing with the rules of inference in Section 2.4, and deferring our discussion of the nonlogical axioms until Section 2.8.

2.3.1 Equality Axioms

We have taken the route of assuming that the equality symbol, =, is a part of the language $\mathcal{L}$. There are three groups of axioms that are designed for this symbol. The first just says that any object is equal to itself:

$$x = x \text{ for each variable } x. \tag{E1}$$

For the second group of axioms, assume that $x_1, x_2, \ldots, x_n$ are variables, $y_1, y_2, \ldots, y_n$ are variables, and f is an n-ary function symbol.

$$\begin{aligned}\left[(x_1 = y_1) \land (x_2 = y_2) \land \cdots \land (x_n = y_n)\right] \rightarrow \\ (f(x_1, x_2, \ldots, x_n) = f(y_1, y_2, \ldots, y_n)).\end{aligned} \tag{E2}$$

The assumptions for the third group of axioms is the same as for the second group, except that R is assumed to be an n-ary relation symbol (R might be the equality symbol, which is seen as a binary relation symbol).

$$\begin{aligned}\left[(x_1 = y_1) \land (x_2 = y_2) \land \cdots \land (x_n = y_n)\right] \rightarrow \\ (R(x_1, x_2, \ldots, x_n) \rightarrow R(y_1, y_2, \ldots, y_n)).\end{aligned} \tag{E3}$$

Axioms (E2) and (E3) are axioms that are designed to allow substitution of equals for equals. Nothing fancier than that.

2.3.2 Quantifier Axioms

The quantifier axioms are designed to allow a very reasonable sort of entry in a deduction. Suppose that we know $\forall x P(x)$. Then, if t is any term of the language, we should be able to state $P(t)$. To avoid problems of the sort outlined at the beginning of Section 1.8, we will demand that the term t be substitutable for the variable x.

$$(\forall x \phi) \rightarrow \phi_t^x, \text{ if } t \text{ is substitutable for } x \text{ in } \phi. \tag{Q1}$$

$$\phi_t^x \rightarrow (\exists x \phi), \text{ if } t \text{ is substitutable for } x \text{ in } \phi. \tag{Q2}$$

In many logic texts, axiom (Q1) would be called universal instantiation, while (Q2) would be known as existential generalization. We will avoid this impressive language and stick with the more mundane (Q1) and (Q2).

2.3.3 Recap

Just to gather all of the logical axioms together in one place, let me state them once again. The set Λ of logical axioms is the collection of all formulas that fall into one of the following categories:

$$x = x \text{ for each variable } x. \tag{E1}$$

$$[(x_1 = y_1) \land (x_2 = y_2) \land \cdots \land (x_n = y_n)] \rightarrow (f(x_1, x_2, \ldots, x_n) = f(y_1, y_2, \ldots, y_n)). \tag{E2}$$

$$[(x_1 = y_1) \land (x_2 = y_2) \land \cdots \land (x_n = y_n)] \rightarrow (R(x_1, x_2, \ldots, x_n) \rightarrow R(y_1, y_2, \ldots, y_n)). \tag{E3}$$

$$(\forall x \phi) \rightarrow \phi_t^x, \text{ if } t \text{ is substitutable for } x \text{ in } \phi. \tag{Q1}$$

$$\phi_t^x \rightarrow (\exists x \phi), \text{ if } t \text{ is substitutable for } x \text{ in } \phi. \tag{Q2}$$

Notice that Λ is decidable: We could write a computer program which, given a formula ϕ, can decide in a finite amount of time whether or not ϕ is an element of Λ.

2.4 Rules of Inference

Having established our set Λ of logical axioms, we must now fix our rules of inference. There will be two types of rules, one dealing with propositional consequence and one dealing with quantifiers.

2.4.1 Propositional Consequence

In all likelihood you are familiar with tautologies of propositional logic. They are simply formulas like $(A \rightarrow B) \leftrightarrow (\neg B \rightarrow \neg A)$. If you are comfortable with tautologies, feel free to skip over the next couple of paragraphs. If not, what follows is a very brief review of a portion of propositional logic.

We work with a restricted language $\mathcal{P}$, consisting only of a set of propositional variables $A, B, C, \ldots$ and the connectives $\vee$ and $\neg$. Notice there are no quantifiers, no relation symbols, no function symbols, and no constants. Formulas of propositional logic are defined as being the collection of all ϕ such that either ϕ is a propositional variable, or ϕ is $(\neg\alpha)$, or ϕ is $(\alpha \vee \beta)$, with α and β being formulas of propositional logic.

Each propositional variable can be assigned one of two truth values, T or F, corresponding to truth and falsity. Given such an assignment (which is really a function v : propositional variables $\rightarrow \{T, F\}$), we can extend v to a function $\overline{v}$ assigning a truth value to any propositional formula as follows:

$$\overline{v}(\phi) = \begin{cases} v(\phi) & \text{if } \phi \text{ is a propositional variable} \\ F & \text{if } \phi \text{ is } (\neg\alpha) \text{ and } \overline{v}(\alpha) = T \\ F & \text{if } \phi \text{ is } (\alpha \vee \beta) \text{ and } \overline{v}(\alpha) = \overline{v}(\beta) = F \\ T & \text{otherwise.} \end{cases}$$

Now we say that a propositional formula ϕ is a tautology if and only if $\overline{v}(\phi) = T$ for any truth assignment v.

One way that you can check whether a given ϕ is a tautology is by constructing a truth table with one row for each possible combination of truth values for the propositional variables that occur in ϕ. Then you fill in the truth table and see whether the truth value associated with the main connective is always true. For example, consider the propositional formula $A \rightarrow (B \rightarrow A)$, which is translated to $\neg A \vee (\neg B \vee A)$. The truth table verifying that this formula is a tautology is

A	B	$\neg A$	$\vee$	$(\neg B$	$\vee$	$A)$
T	T	F	T	F	T	T
T	F	F	T	T	T	T
F	T	T	T	F	F	F
F	F	T	T	T	T	F

To discuss propositional consequence in first-order logic, we will transfer our formulas to the realm of propositional logic and use the idea of tautology in that area. To be specific, given β, an $\mathcal{L}$-formula of first-order logic, here is a procedure that will convert β to a formula β_P of propositional logic corresponding to β:

1. Find all subformulas of β of the form $\forall x \alpha$ that are not in the scope of another quantifier. Replace them with propositional variables in a systematic fashion. This means that if $\forall y Q(y, c)$ appears twice in β, it is replaced by the same letter both times, and distinct subformulas are replaced with distinct letters.

2. Find all atomic formulas that remain, and replace them systematically with new propositional variables.

3. At this point, β will have been replaced with a propositional formula β_P.

For example, suppose that we look at the $\mathcal{L}$-formula

$$(\forall x P(x) \wedge Q(c, z)) \rightarrow (Q(c, z) \vee \forall x P(x)).$$

For the first step of the procedure above, we replace the quantified subformulas with the propositional letter B:

$$(B \wedge Q(c, z)) \rightarrow (Q(c, z) \vee B).$$

To finish the transformation to a propositional formula, replace the atomic formula with a propositional letter:

$$(B \wedge A) \rightarrow (A \vee B).$$

Notice that if β_P is a tautology, then β is valid, but the converse of this statement fails. For example, if β is

$$\left[(\forall x)(\theta) \wedge (\forall x)(\theta \rightarrow \rho)\right] \rightarrow (\forall x)(\rho),$$

then β is valid, but β_P would be $[A \land B] \to C$, which is certainly not a tautology.

We are now almost at a point where we can state our propositional rule of inference. Recall that a rule of inference is an ordered pair $\langle \Gamma, \phi \rangle$, where Γ is a set of $\mathcal{L}$-formulas and ϕ is an $\mathcal{L}$-formula.

Definition 2.4.1. Suppose that Γ_P is a set of propositional formulas and ϕ_P is a propositional formula. We will say that ϕ_P is a **propositional consequence of** Γ_P if every truth assignment that makes each propositional formula in Γ_P true also makes ϕ_P true. Notice that ϕ_P is a tautology if and only if ϕ_P is a propositional consequence of $\emptyset$.

Lemma 2.4.2. *If $\Gamma_P = \{\gamma_{1P}, \gamma_{2P}, \ldots, \gamma_{nP}\}$ is a nonempty finite set of propositional formulas and ϕ_P is a propositional formula, then ϕ_P is a propositional consequence of Γ_P if and only if*

$$[\gamma_{1P} \land \gamma_{2P} \land \cdots \land \gamma_{nP}] \to \phi_P$$

is a tautology.

Proof: Exercise 3. ■

Now we extend our definition of propositional consequence to include formulas of first-order logic:

Definition 2.4.3. Suppose that Γ is a finite set of $\mathcal{L}$-formulas and ϕ is an $\mathcal{L}$-formula. We will say that ϕ is a **propositional consequence of** Γ if ϕ_P is a propositional consequence of Γ_P, where ϕ_P and Γ_P are the results of applying the procedure on the preceding page uniformly to ϕ and all of the formulas in Γ.

Example 2.4.4. Suppose that $\mathcal{L}$ contains two unary relation symbols, P and Q. Let Γ be the set

$$\{\forall x P(x) \to \exists y Q(y), \exists y Q(y) \to P(x), \neg P(x) \leftrightarrow (y = z)\}.$$

If we let ϕ be the formula $\forall x P(x) \to \neg(y = z)$, then by applying our procedure uniformly to the elements of Γ and ϕ, we see that

$$\Gamma_P \text{ is } \{A \to B, B \to C, \neg C \leftrightarrow D\}$$

and ϕ_P is $A \to \neg D$, where the fact that we have substituted the same propositional variables for the same formulas in ϕ and the elements of Γ is ensured by our applying the procedure *uniformly* to all of the formulas in question. At this point it is easy to verify that ϕ is a propositional consequence of Γ.

Finally, our rule of inference:

Definition 2.4.5. If Γ is a finite set of $\mathcal{L}$-formulas, ϕ is an $\mathcal{L}$-formula, and ϕ is a propositional consequence of Γ, then $\langle \Gamma, \phi \rangle$ is a **rule of inference of type (PC)**.

> *Chaff:* All of this formalism just might be hiding what is really going on here. What rule (PC) says is that if you have proved γ_1 and γ_2 and $[(\gamma_1 \land \gamma_2) \to \phi]_P$ is a tautology, then you may conclude ϕ. Nothing fancier than that.
>
> Also notice that if ϕ is a formula such that ϕ_P is a tautology, rule (PC) allows us to assert ϕ in any deduction, using $\Gamma = \emptyset$.

2.4.2 Quantifier Rules

The motivation behind our quantifier rules is very simple. Suppose, without making any particular assumptions about x, that you were able to prove "x is an ambitious aardvark." Then it seems reasonable to claim that you have proved "$(\forall x)x$ is an ambitious aardvark." Dually, if you were able to prove the Riemann Hypothesis from the assumption that "x is a bossy bullfrog," then from the assumption "$(\exists x)x$ is a bossy bullfrog," you should still be able to prove the Riemann Hypothesis.

Definition 2.4.6. Suppose that the variable x is not free in the formula ψ. Then both of the following are **rules of inference of type (QR)**:

$$\langle \{\psi \to \phi\}, \psi \to (\forall x \phi) \rangle$$
$$\langle \{\phi \to \psi\}, (\exists x \phi) \to \psi \rangle.$$

The "not making any particular assumptions about x" comment is made formal by the requirement that x not be free in ψ.

> *Chaff:* Just to make sure that you are not lost in the brackets of the definition, what we are saying here is that if x is not free in ψ:

1. From the formula $\psi \to \phi$, you may deduce $\psi \to (\forall x \phi)$.
2. From the formula $\phi \to \psi$, you may deduce $(\exists x \phi) \to \psi$.

2.4.3 Exercises

1. We claim that the collection Λ of logical axioms is decidable. Outline an algorithm which, given an $\mathcal{L}$-formula θ, outputs "yes" if θ is an element of Λ and outputs "no" if θ is not an element of Λ. You do not have to be too fussy. Notice that you have to be able to decide if a term t is substitutable for a variable x is a formula ϕ. See Exercise 7 in Section 1.8.1.

2. Show that the set of rules of inference is decidable. So outline an algorithm that will decide, given a finite set of formulas Γ and a formula θ, whether or not $\langle \Gamma, \theta \rangle$ is a rule of inference.

3. Prove Lemma 2.4.2.

4. Write a deduction of the second quantifier axiom (Q2) (on page 57) without using (Q2) as an axiom.

5. For each of the following, decide if ϕ is a propositional consequence of Γ and justify your assertion.

(a) Γ is $\{(\forall x P(x)) \to Q(y), (\forall x P(x)) \lor (\forall x R(x)), \exists x \neg R(x)\}$; ϕ is $Q(y)$.

(b) Γ is $\{x = y \land Q(y), Q(y) \lor x + y < z\}$; ϕ is $x + y < z$.

(c) Γ is $\{P(x, y, x), x < y \lor M(w, p), (\neg P(x, y, x)) \land (\neg x < y)\}$; ϕ is $\neg M(w, p)$.

6. Prove that if θ is not valid, then θ_P is not a tautology. Deduce that if θ_P is a tautology, then θ is valid.

2.5 Soundness

Mathematicians are by nature a conservative bunch. I speak not of political or social leanings, but of their professional outlook. In particular, a mathematician likes to know that when something has been proved, it is true. In this section we will prove a theorem that shows that the logical system that we have developed has this highly desirable property. This result is called the Soundness Theorem.

Let me restate the list of requirements that we set out on page 49 for our axioms and rules of inference:

1. There will be an algorithm that will decide, given a formula θ, whether or not θ is a logical axiom.

2. There will be an algorithm that will decide, given a finite set of formulas Γ and a formula θ, whether or not $\langle \Gamma, \theta \rangle$ is a rule of inference.

3. For each rule of inference $\langle \Gamma, \theta \rangle$, Γ will be a finite set of formulas.

4. Each logical axiom will be valid.

5. Our rules of inference will preserve truth. In other words, for each rule of inference $\langle \Gamma, \theta \rangle$, $\Gamma \models \theta$.

These requirements serve two purposes: They allow us to verify mechanically that an alleged deduction is in fact a deduction, and they provide the basis of the Soundness Theorem. Of course, we must first verify that the system of axioms and rules that we have set out in the preceding two sections satisfy these requirements.

That the first three requirements above are satisfied by our deduction system was noted as the axioms and rules were presented. These are the rules that are needed for deduction verification. We will discuss the last two requirements in more detail and then use those requirements to prove the Soundness Theorem.

Theorem 2.5.1. *The logical axioms are valid.*

Proof: We must check both the equality axioms and the quantifier axioms. First, consider equality axioms of type (E2). [(E1) and (E3) will be proved in the Exercises.]

> *Chaff:* Let me mention that we will use Theorem 2.6.2 in this proof. Although the presentation of that result has been delayed in order to aid the flow of the exposition, you may want to look at the statement of that theorem now so you won't be surprised when it appears.

So fix a structure $\mathfrak{A}$ and an assignment function $s : Vars \to A$. We must show that

$$\mathfrak{A} \models \Big([(x_1 = y_1) \land (x_2 = y_2) \land \cdots \land (x_n = y_n)] \rightarrow \\ (f(x_1, x_2, \ldots, x_n) = f(y_1, y_2, \ldots, y_n)) \Big) [s].$$

As the formula in question is an implication, we may assume that the antecedent is satisfied by the pair $\langle \mathfrak{A}, s \rangle$, and thus $s(x_1) = s(y_1), s(x_2) = s(y_2), \ldots,$ and $s(x_n) = s(y_n)$. We must prove that $\mathfrak{A} \models (f(x_1, x_2, \ldots, x_n) = f(y_1, y_2, \ldots, y_n))[s]$. From the definition of satisfaction (Definition 1.7.4), we know this means that we have to show

$$\overline{s}(f(x_1, x_2, \ldots, x_n)) = \overline{s}(f(y_1, y_2, \ldots, y_n)).$$

Now we look at the definition of term assignment function (Definition 1.7.3) and see that we must prove

$$f^{\mathfrak{A}}(\overline{s}(x_1), \overline{s}(x_2), \ldots, \overline{s}(x_n)) = f^{\mathfrak{A}}(\overline{s}(y_1), \overline{s}(y_2), \ldots, \overline{s}(y_n)).$$

But since $\overline{s}(x_i) = s(x_i) = s(y_i) = \overline{s}(y_i)$, and since $f^{\mathfrak{A}}$ is a function, this is true. Thus our equality axiom (E2) is valid.

Now we examine the quantifier axiom of type (Q1), reserving (Q2) for the Exercises. Once again, fix $\mathfrak{A}$ and s, and assume that the term t is substitutable for the variable x in the formula ϕ. We must show that

$$\mathfrak{A} \models \left[(\forall x \phi) \rightarrow \phi^x_t\right][s].$$

So once again, we assume that $\mathfrak{A} \models (\forall x \phi)[s]$, and we show that $\mathfrak{A} \models \phi^x_t[s]$. By assumption, $\mathfrak{A} \models \phi[s[x|a]]$ for any element $a \in A$, so in particular, $\mathfrak{A} \models \phi[s[x|\overline{s}(t)]]$.

Informally, this says that ϕ is true in $\mathfrak{A}$ with assignment function s, where you interpret x as $\overline{s}(t)$. It is plausible, given our assumption that t is substitutable for x in ϕ, that if we altered the formula ϕ by replacing x by t, then ϕ^x_t would be true in $\mathfrak{A}$ with assignment function s. This is the content of Theorem 2.6.2. Since we know that $\mathfrak{A} \models \phi[s[x|\overline{s}(t)]]$ and Theorem 2.6.2 states that this is equivalent to $\mathfrak{A} \models \phi^x_t[s]$, we have established $\mathfrak{A} \models \phi^x_t[s]$, so we have proved that axioms of type (Q1) are valid.

Thus, modulo your proofs of (E1), (E2), and (Q2) and the delayed proof of Theorem 2.6.2, all of our logical axioms are valid, and our proof is complete. ■

This leaves one more item on our list of requirements to check. We must show that our rules of inference preserve truth.

Theorem 2.5.2. *Suppose that $\langle \Gamma, \theta \rangle$ is a rule of inference. Then $\Gamma \models \theta$.*

Proof: First, assume that $\langle \Gamma, \theta \rangle$ is a rule of type (PC). Then Γ is finite, and by Lemma 2.4.2, we know that

$$[\gamma_{1P} \land \gamma_{2P} \land \cdots \land \gamma_{nP}] \rightarrow \theta_P$$

is a tautology, where $\Gamma_P = \{\gamma_{1P}, \gamma_{2P}, \ldots, \gamma_{nP}\}$ is the set of propositional formulas corresponding to Γ and θ_P is the propositional formula corresponding to θ. But then, by Exercise 6 on page 62, we know that

$$[\gamma_1 \land \gamma_2 \land \cdots \land \gamma_n] \rightarrow \theta$$

is valid, and thus $\Gamma \models \theta$.

The other possibility is that our rule of inference is a quantifier rule. So, suppose that x is not free in ψ. We show that $(\psi \rightarrow \phi) \models [\psi \rightarrow (\forall x \phi)]$, leaving the other (QR) rule for the Exercises.

So fix a structure $\mathfrak{A}$ and assume that $\mathfrak{A} \models (\psi \rightarrow \phi)$. Thus our assumption is that for any assignment s, $\mathfrak{A} \models (\psi \rightarrow \phi)[s]$. We must show that $\mathfrak{A} \models (\psi \rightarrow \forall x \phi)$, which means that we must show that $(\psi \rightarrow \forall x \phi)$ is satisfied in $\mathfrak{A}$ under every assignment function. So let an assignment function $t : Vars \rightarrow A$ be given. We must show that $\mathfrak{A} \models (\psi \rightarrow \forall x \phi)[t]$. If $\mathfrak{A} \not\models \psi[t]$, we are done, so assume that $\mathfrak{A} \models \psi[t]$. We want to prove that $\mathfrak{A} \models \forall x \phi[t]$, which means that if a is any element of A, we must show that $\mathfrak{A} \models \phi[t[x|a]]$.

We know, by assumption, that $\mathfrak{A} \models (\psi \rightarrow \phi)[t[x|a]]$. Furthermore, Proposition 1.7.7 tells us that $\mathfrak{A} \models \psi[t[x|a]]$, as $\mathfrak{A} \models \psi[t]$, and t and $t[x|a]$ agree on all of the free variables of ψ (x is not free in ψ by assumption). But then, by the definition of satisfaction, $\mathfrak{A} \models \phi[t[x|a]]$, and we are finished. ■

We are now at a point where we can prove the Soundness Theorem. The idea behind this theorem is very simple. Suppose that Σ is a set of $\mathcal{L}$-formulas and suppose that there is a deduction of ϕ from Σ. What the Soundness Theorem tells us is that in any structure $\mathfrak{A}$ that makes all of the formulas of Σ true, ϕ is true as well.

Theorem 2.5.3 (Soundness). *If* $\Sigma \vdash \phi$, *then* $\Sigma \models \phi$.

Proof: Let $\text{Thm}_\Sigma = \{\phi \mid \Sigma \vdash \phi\}$, and let $C = \{\phi \mid \Sigma \models \phi\}$. We show that $\text{Thm}_\Sigma \subseteq C$, which proves the theorem.

Notice that C has the following characteristics:

1. $\Sigma \subseteq C$. If $\sigma \in \Sigma$, then certainly $\Sigma \models \sigma$.

2. $\Lambda \subseteq C$. As the logical axioms are valid, they are true in any structure. Thus $\Sigma \models \lambda$ for any logical axiom λ, which means that if $\lambda \in \Lambda$, then $\lambda \in C$, as needed.

3. If $\langle \Gamma, \theta \rangle$ is a rule of inference and $\Gamma \subseteq C$, then $\theta \in C$. So assume that $\Gamma \subseteq C$. To prove $\theta \in C$ we must show that $\Sigma \models \theta$. Fix a structure $\mathfrak{A}$ such that $\mathfrak{A} \models \Sigma$. We must prove that $\mathfrak{A} \models \theta$.

   If γ is any element of Γ, then since $\gamma \in C$, we know that $\Sigma \models \gamma$. Since $\mathfrak{A} \models \Sigma$ and $\Sigma \models \gamma$, we know that $\mathfrak{A} \models \gamma$. But this says that $\mathfrak{A} \models \gamma$ for each $\gamma \in \Gamma$, so $\mathfrak{A} \models \Gamma$. But Theorem 2.5.2 tells us that $\Gamma \models \theta$, since $\langle \Gamma, \theta \rangle$ is a rule of inference. Therefore, since $\mathfrak{A} \models \Gamma$ and $\Gamma \models \theta$, $\mathfrak{A} \models \theta$, as needed.

So C is a set of the type outlined in Proposition 2.2.4, and by that proposition, $\text{Thm}_\Sigma \subseteq C$, as needed. ■

Notice that the Soundness Theorem begins to tie together the notions of deducibility and logical implication. It says, "If there is a deduction from Σ of ϕ, then Σ logically implies ϕ." Thus the purely syntactic notion of deduction, a notion that relies only upon typographical considerations, is linked to the notions of truth and logical implication, ideas that are inextricably tied to mathematical structures and their properties. This linkage will be tightened in Chapter 3.

> *Chaff:* The proof of the Soundness Theorem that I have presented above has the desirable qualities of being neat and quick. It emphasizes a core fact about the consequences of Σ, namely that Thm_Σ is the smallest set of formulas satisfying the given three conditions. Unfortunately, the proof has the less desirable attribute of being

pretty abstract. Exercise 5 outlines a more direct, less abstract proof of the Soundness Theorem.

2.5.1 Exercises

1. Ingrid walks into your office one day and announces that she is puzzled. She has a set of axioms Σ in the language of number theory, and she has a formula ϕ that she has proved using the assumptions in Σ. Unfortunately, ϕ is a statement that is not true in the standard model $\mathfrak{N}$. Is this a problem? If it is a problem, what possible explanations can you think of that would explain what went wrong? If it is not a problem, *why* is it not a problem?

2. Prove that the equality axioms of type (E1) and (E3) are valid.

3. Show that the quantifier axiom of type (Q2) is valid.

4. Show that, if x is not free in ψ, $(\phi \rightarrow \psi) \models [(\exists x \phi) \rightarrow \psi]$.

5. Prove the Soundness Theorem by induction on the complexity of the proof of ϕ. For the base cases, ϕ is either a logical axiom or a member of Σ. Then assume that ϕ is proved by reference to a rule of inference show that in this case as well, $\Sigma \models \phi$.

2.6 Two Technical Lemmas

In this section we present two rather technical lemmas that we need to complete the proof of Theorem 2.5.1. The proofs that are involved are not pretty, and if you are the trusting sort, you may want to scan through this section rather quickly. On the other hand, if you come to grips with these results, you will gain a better appreciation for the details of substitutability and assignment functions.

To motivate the first lemma, consider this example: Suppose that we are working in the language of number theory and that the structure under consideration comprises the natural numbers. Let the term u be $x \cdot v$ and the term t be $y + z$. Then u_t^x is $(y + z) \cdot v$. Now we have to fix a couple of assignment functions. Let the assignment function s look like this:

Vars	s
x	12
y	3
z	7
v	4
$\vdots$	$\vdots$

So $s(x) = 12$, $s(y) = 3$, and so on.

Now, suppose that s' is an assignment function that is just like s, except that s' sends x to the value $\overline{s}(t)$, which is $\overline{s}(y+z) = 3+7 = 10$:

Vars	s	s'
x	12	10
y	3	3
z	7	7
v	4	4
$\vdots$	$\vdots$	$\vdots$

Now, if you compare $\overline{s}(u_t^x)$ and $\overline{s'}(u)$, you find that

$$\overline{s}(u_t^x) = \overline{s}((y + z) \cdot v) = (3 + 7) \cdot 4 = 10 \cdot 4 = 40$$

$$\overline{s'}(u) = \overline{s'}(x \cdot v) = 10 \cdot 4 = 40.$$

So, in this situation, the element of the universe that is assigned by s to u_t^x is the same as the element of the universe that is assigned by s' to u. In some sense, the lemma states that it does not matter whether you alter the term or the assignment function, the result is the same.

Here is the formal statement:

Lemma 2.6.1. *Suppose that u is a term, x is a variable, and t is a term. Suppose that $s : Vars \to A$ is a variable assignment function and that $s' = s[x|\overline{s}(t)]$. Then $\overline{s}(u_t^x) = \overline{s'}(u)$.*

Proof: The proof is by induction on the complexity of the term u. If u is a variable and $u = x$, then

$$\begin{aligned} \overline{s}(u_t^x) &= \overline{s}(x_t^x) \\ &= \overline{s}(t) \\ &= \overline{s'}(x) \\ &= \overline{s'}(u). \end{aligned}$$

If u is a variable and $u = y \neq x$, then

$$\begin{aligned}\overline{s}(u_t^x) &= \overline{s}(y_t^x) \\ &= \overline{s}(y) \\ &= s(y) \\ &= s'(y) \\ &= \overline{s'}(u).\end{aligned}$$

If u is a constant symbol c, then $\overline{s}(u_t^x) = \overline{s}(c_t^x) = \overline{s}(c) = c^{\mathfrak{A}} = \overline{s'}(u)$.

The last inductive case is if u is $f(r_1, r_2, \ldots, r_n)$, with each r_i a term. In this case,

$$\begin{aligned}\overline{s}(u_t^x) &= \overline{s}([f(r_1, r_2, \ldots, r_n)]_t^x) \\ &= \overline{s}\left(f\left((r_1)_t^x, (r_2)_t^x, \ldots, (r_n)_t^x\right)\right) \\ &= f^{\mathfrak{A}}(\overline{s}[(r_1)_t^x], \overline{s}[(r_2)_t^x], \ldots, \overline{s}[(r_n)_t^x]) && \text{definition of } \overline{s} \\ &= f^{\mathfrak{A}}(\overline{s'}(r_1), \overline{s'}(r_2), \ldots, \overline{s'}(r_n)) && \text{inductive hypothesis} \\ &= \overline{s'}(f(r_1, r_2, \ldots, r_n)) && \text{definition of } \overline{s'} \\ &= \overline{s'}(u).\end{aligned}$$

So for every term u, $\overline{s}(u_t^x) = \overline{s'}(u)$. ■

Chaff: That was hard. If you understood that proof the first time through, you have done something quite out of the ordinary. If, on the other hand, you are a mere mortal, you might want to work through the proof again, keeping an example in mind as you work. Pick a language, terms u and t in your language, and a variable x. Fix a particular assignment function s. Then just follow through the steps of the proof, keeping track of where everything goes. I would write it out for you, but you will get more out of doing it for yourself. Go to it!

Our next technical result is the lemma that we quoted explicitly in the proof of Theorem 2.5.1. This theorem states that as long as t is substitutable for x in ϕ, the two different ways of evaluating the truth of "ϕ, where you interpret x as t" coincide. The first way of evaluating the truth would be by forming the formula ϕ_t^x and seeing if $\mathfrak{A} \models \phi_t^x[s]$. The second way would be to change the assignment function s to interpret x as $\overline{s}(t)$ and checking whether the original formula ϕ is true with this new assignment function. The theorem states that the two methods are equivalent.

Theorem 2.6.2. *Suppose that ϕ is an $\mathcal{L}$-formula, x is a variable, t is a term, and t is substitutable for x in ϕ. Suppose that $s : Vars \to A$ is a variable assignment function and that $s' = s[x|\overline{s}(t)]$. Then $\mathfrak{A} \models \phi_t^x[s]$ if and only if $\mathfrak{A} \models \phi[s']$.*

Proof: We use induction on the complexity of ϕ. The first base case is where ϕ is $u_1 = u_2$, where u_1 and u_2 are terms. Then the following are equivalent:

$$\begin{array}{ll} \mathfrak{A} \models \phi_t^x[s] & \\ \mathfrak{A} \models (u_1)_t^x = (u_2)_t^x[s] & \\ \overline{s}\big((u_1)_t^x\big) = \overline{s}\big((u_2)_t^x\big) & \text{definition of satisfaction} \\ \overline{s'}(u_1) = \overline{s'}(u_2) & \text{by Lemma 2.6.1} \\ \mathfrak{A} \models \phi[s'] & \end{array}$$

The second base case is where ϕ is $R(u_1, u_2, \ldots, u_n)$. This case is similar to the case above.

The inductive cases involving the connectives $\vee$ and $\neg$ follow immediately from the inductive hypothesis.

This leaves the last inductive case, where ϕ is $\forall y \psi$. We break this case down into two subcases: In the first subcase x is y, and in the second subcase x is not y.

If ϕ is $\forall y \psi$ and y is x, then ϕ_t^x is ϕ. Therefore, $\mathfrak{A} \models \phi_t^x[s]$ if and only if $\mathfrak{A} \models \phi[s]$. But as s and s' agree on all of the free variables of ϕ (x is not free), by Proposition 1.7.7, $\mathfrak{A} \models \phi[s]$ if and only if $\mathfrak{A} \models \phi[s']$, as needed for this subcase.

The second subcase, where ϕ is $\forall y \psi$ and y is not x, is examined in two sub-subcases:

Sub-subcase 1: If ϕ is $\forall y \psi$, y is not x, and x is not free in ψ, then we know by Exercise 5 in Section 1.8.1 that ψ_t^x is ψ, and thus ϕ_t^x is ϕ. But then

$$\begin{array}{ll} \mathfrak{A} \models \phi_t^x[s] & \text{iff} \\ \mathfrak{A} \models \phi[s] & \text{iff} \\ \mathfrak{A} \models \phi[s'], & \end{array}$$

as s and s' agree on the free variables of ϕ.

Sub-subcase 2: If ϕ is $\forall y\psi$, y is not x, and x *is* free in ψ, then as t is substitutable for x in ϕ (we had to use that assumption somewhere, didn't we?), we know that y does not occur in t and t is substitutable for x in ψ. Then we have

$$\begin{aligned} &\mathfrak{A} \models \phi_t^x[s] && \text{iff} \\ &\mathfrak{A} \models (\forall y)(\psi_t^x)[s] && \text{iff} \\ &\mathfrak{A} \models (\psi_t^x)[s[y|a]] && \text{for every } a \in A. \end{aligned}$$

But we also know that

$$\begin{aligned} &\mathfrak{A} \models \phi[s'] && \text{iff} \\ &\mathfrak{A} \models (\forall y)(\psi)[s'] && \text{iff} \\ &\mathfrak{A} \models \psi[s'[y|a]] && \text{for every } a \in A. \end{aligned}$$

But since x is not y, we know that for any $a \in A$, $s'[y|a] = s[y|a][x|\bar{s}(t)]$, so by the inductive hypothesis (notice that t is substitutable for x in ψ) we have

$$\mathfrak{A} \models (\psi_t^x)[s[y|a]] \text{ iff } \mathfrak{A} \models \psi[s'[y|a]].$$

So $\mathfrak{A} \models \phi_t^x[s]$ if and only if $\mathfrak{A} \models \phi[s']$, as needed. ■

2.7 Properties of Our Deductive System

Having gone through all the trouble of setting out our deductive system, we will now prove a few things both in and about that system. First, we will show that we can prove, in our deductive system, that equality is an equivalence relation.

Theorem 2.7.1.

1. $\vdash x = x$.

2. $\vdash x = y \rightarrow y = x$.

3. $\vdash (x = y \land y = z) \rightarrow x = z$.

Proof: We show that we can find deductions establishing that $=$ is reflexive, symmetric, and transitive in turn.

1. This is a logical axiom of type (E1).

2. Here is the needed deduction. Notice that the notations off to the right are listed only as an aid to the reader.

$$[x = y \land x = x] \to [x = x \to y = x] \qquad \text{(E3)}$$
$$x = x \qquad \text{(E1)}$$
$$x = y \to y = x. \qquad \text{(PC)}$$

3. Again, we present a deduction:

$$[x = x \land y = z] \to [x = y \to x = z] \qquad \text{(E3)}$$
$$x = x \qquad \text{(E1)}$$
$$(x = y \land y = z) \to x = z. \qquad \text{(PC)} \quad \blacksquare$$

Chaff: Notice that we have done a bit more than prove that equality is an equivalence relation. (Heck, you've known *that* since fourth grade.) Rather, we've shown that our deductive system, with the axioms and rules of inference that have been outlined in this chapter, is powerful enough to *prove* that equality is an equivalence relation. There will be a fair bit of "our deductive system is strong enough to do such-and-such" in the pages to come.

We now prove some general properties of our deductive system. We start off with a lemma that seems somewhat problematical, but it will help us to think a little more carefully about what our deductions do for us.

Lemma 2.7.2. *$\Sigma \vdash \theta$ if and only if $\Sigma \vdash \forall x\theta$.*

***Proof*:** First, suppose that $\Sigma \vdash \theta$. Here is a deduction from Σ of $\forall x\theta$:

$$\vdots \qquad \text{Deduction of } \theta$$
$$\theta$$
$$\big[(\forall y(y = y)) \lor \neg(\forall y(y = y))\big] \to \theta \qquad \text{(PC)}$$
$$\big[(\forall y(y = y)) \lor \neg(\forall y(y = y))\big] \to (\forall x\theta) \qquad \text{(QR)}$$
$$\forall x\theta. \qquad \text{(PC)}$$

There are a couple of things to point out about this proof. The first use of (PC) is justified by the fact that if θ is true, then (anything $\rightarrow \theta$) is also true. The second use of (PC) depends on the fact that $\left[(\forall y(y=y)) \vee \neg(\forall y(y=y))\right]$ is a tautology, and thus $\forall x\theta$ is a propositional consequence of the implication. As for the (QR) step of the deduction, notice that the variable x is not free in the sentence $\left[(\forall y(y=y)) \vee \neg(\forall y(y=y))\right]$, making the use of the quantifier rule legitimate.

Now, suppose that $\Sigma \vdash \forall x\theta$. Here is a deduction from Σ of θ (recall that θ_x^x is θ):

$\vdots$	Deduction of $\forall x\theta$
$\forall x\theta$	
$\forall x\theta \rightarrow \theta_x^x$	(Q1)
θ_x^x.	(PC)

Thus $\Sigma \vdash \theta$ if and only if $\Sigma \vdash \forall x\theta$. ■

Here is an example to show how strange this lemma might seem. Suppose that Σ consists of the single formula $x = \overline{5}$. Then certainly $\Sigma \vdash x = \overline{5}$, and so, by the lemma, $\Sigma \vdash (\forall x)(x = \overline{5})$. You might be tempted to say that by assuming x was equal to five, we have proved that *everything* is equal to five. But that is not quite what is going on. If $x = \overline{5}$ is true in a model $\mathfrak{A}$, that means that $\mathfrak{A} \models x = \overline{5}[s]$ for every assignment function s. And since for every $a \in A$, there is an assignment function s such that $s(x) = a$, it must be true that every element of A is equal to 5, so the universe A has only one element, and everything *is* equal to 5. So our deduction of $(\forall x)(x = \overline{5})$ has preserved truth, but our assumption was much stronger than it appeared at first glance. And the moral of our story is: For a formula to be true in a structure, it must be satisfied in that structure with *every* assignment function.

Lemma 2.7.3. *Suppose that $\Sigma \vdash \theta$. Then if Σ' is formed by taking any $\sigma \in \Sigma$ and adding or deleting a universal quantifier whose scope is the entire formula, $\Sigma' \vdash \theta$.*

Proof: This follows immediately from Lemma 2.7.2. Suppose that $\forall x\sigma$ is in Σ'. By the preceding, $\Sigma' \vdash \sigma$. Then, given a deduction from Σ of θ, to produce a deduction from Σ' of θ, first write down a deduction from Σ' of σ, and then copy your deduction from Σ of θ. Having already established σ, this deduction will be a valid deduction from Σ'.

The proof in the case that $\forall x \sigma$ is an element of Σ and it is replaced by σ in Σ' is analogous. ■

Notice that one consequence of this lemma is the fact that if we know $\Sigma \vdash \theta$, we can assume (if we like) that every element of Σ is a sentence: By quoting Lemma 2.7.3 several times, we can replace each $\sigma \in \Sigma$ with its universal closure.

Now we will show that in at least some sense, the system of deductions that we have developed mirrors the process that mathematicians use to prove theorems. Suppose you were asked to prove the theorem: *If A is a square, then A is a rectangle.* A perfectly reasonable way to attack this theorem would be to assume that A is a square, and using that assumption, prove that A is a rectangle. But notice that you have not been asked to prove that A is a rectangle. You were asked to prove an implication! The Deduction Theorem says that there is a deduction of ϕ from the assumption θ if and only if there is a deduction of the implication $\theta \rightarrow \phi$. (A bit of notation: Rather than writing the formally correct $\Sigma \cup \{\theta\} \vdash \phi$, we shall omit the braces and write $\Sigma \cup \theta \vdash \phi$.)

Theorem 2.7.4 (The Deduction Theorem). *Suppose that θ is a sentence and Σ is a set of formulas. Then $\Sigma \cup \theta \vdash \phi$ if and only if $\Sigma \vdash (\theta \rightarrow \phi)$.*

Proof: First, suppose that $\Sigma \vdash (\theta \rightarrow \phi)$. Then, as the same deduction would show that $\Sigma \cup \theta \vdash (\theta \rightarrow \phi)$, and as $\Sigma \cup \theta \vdash \theta$ by a one-line deduction, and as ϕ is a propositional consequence of θ and $(\theta \rightarrow \phi)$, we know that $\Sigma \cup \theta \vdash \phi$.

For the more difficult direction we will make use of Proposition 2.2.4. Suppose that $C = \{\phi \mid \Sigma \vdash (\theta \rightarrow \phi)\}$. If we show that C contains $\Sigma \cup \theta$, C contains all the axioms of Λ, and C is closed under the rules of inference as noted in Proposition 2.2.4, then by that proposition we will know that $\{\phi \mid \Sigma \cup \theta \vdash \phi\} \subseteq C$. In other words, we will know that if $\Sigma \cup \theta \vdash \phi$, then $\Sigma \vdash (\theta \rightarrow \phi)$, which is what we need to show.

So it remains to prove that C has the properties listed in the preceding paragraph.

1. $\Sigma \subseteq C$: If $\sigma \in \Sigma$, then $\Sigma \vdash \sigma$. But then $\Sigma \vdash (\theta \rightarrow \sigma)$, as this is a propositional consequence of σ.
2. $\theta \in C$: $\Sigma \vdash \theta \rightarrow \theta$, as this is a tautology.
3. $\Lambda \subseteq C$: This is identical to (1).
4. C is closed under the rules:
   (a) *Rule (PC):* Suppose that $\gamma_1, \gamma_2, \ldots, \gamma_n$ are all elements of C and ϕ is a propositional consequence of $\{\gamma_1, \gamma_2, \ldots, \gamma_n\}$. We must show that $\phi \in C$. By assumption, $\Sigma \vdash (\theta \rightarrow \gamma_1)$, $\Sigma \vdash (\theta \rightarrow \gamma_2)$, ... , $\Sigma \vdash (\theta \rightarrow \gamma_n)$. But then as $(\theta \rightarrow \phi)$ is a propositional consequence of the set

   $$\{(\theta \rightarrow \gamma_1), (\theta \rightarrow \gamma_2), \ldots, (\theta \rightarrow \gamma_n)\},$$

   we have that $\Sigma \vdash (\theta \rightarrow \phi)$. In other words, $\phi \in C$, as needed.

   (b) *Quantifier Rules:* Suppose that $\psi \rightarrow \phi$ is in C and x is not free in ψ. We want to show that $(\psi \rightarrow \forall x \phi)$ is an element of C. In other words, we have to show that

   $$\Sigma \vdash \left[\theta \rightarrow (\psi \rightarrow \forall x \phi)\right].$$

   By assumption we have

$\Sigma \vdash \left[\theta \rightarrow (\psi \rightarrow \phi)\right]$	$\psi \rightarrow \phi$ is in C
$\Sigma \vdash (\theta \wedge \psi) \rightarrow \phi$	propositional consequence
$\Sigma \vdash (\theta \wedge \psi) \rightarrow \forall x \phi$	rule (QR)
$\Sigma \vdash \left[\theta \rightarrow (\psi \rightarrow \forall x \phi)\right]$	propositional consequence

   Notice that our use of rule (QR) is legitimate since we know that θ is a sentence, so x is not free in θ. But the last line of our argument says that $(\psi \rightarrow \forall x \phi) \in C$, which is what we needed to show.

   The other quantifier rule, dealing with the existential quantifier, is proved similarly.

So we have shown that C contains θ, all the elements of Σ and Λ, and C is closed under the rules. This finishes the proof of the Deduction Theorem. ∎

2.7.1 Exercises

1. Lemma 2.7.2 tells us that $\Sigma \vdash \theta$ if and only if $\Sigma \vdash \forall x\theta$. What happens if we replace the universal quantifier by an existential quantifier? So suppose that $\Sigma \vdash \theta$. Must $\Sigma \vdash \exists x\theta$? Now assume that $\Sigma \vdash \exists x\theta$. Does Σ necessarily prove θ?

2. Finish the proof of Lemma 2.7.3 by considering the case when $\forall x\sigma$ is an element of Σ and is replaced by σ in Σ'.

3. Many authors demand that axioms be sentences rather than formulas. Explain how Lemma 2.7.2 implies that we could replace all of our axioms by their universal closures without changing the strength of our deductive system.

4. Suppose that η is a sentence. Prove that $\Sigma \vdash \eta$ if and only if $\Sigma \cup (\neg\eta) \vdash \big[(\forall x)x = x\big] \wedge \neg\big[(\forall x)x = x\big]$. Notice that this exercise tells us that our deductive system allows us to do proofs by contradiction.

5. Suppose that P is a unary relation symbol and show that

$$\vdash \big[(\forall x)P(x)\big] \rightarrow \big[(\exists x)P(x)\big].$$

[*Suggestion:* Proof by contradiction (see Exercise 4) works nicely here.]

6. If P is a binary relation symbol, show that

$$(\forall x)(\forall y)P(x, y) \vdash (\forall y)(\forall z)P(z, y).$$

7. Let P and Q be unary relation symbols, and show that

$$\vdash [(\forall x)(P(x)) \wedge (\forall x)(Q(x))] \rightarrow (\forall x)\,[P(x) \wedge Q(x)]\,.$$

2.8 Nonlogical Axioms

When we are trying to prove theorems in mathematics, there are almost always additional axioms, beyond the set of logical axioms Λ, that we use. If we are trying to prove a theorem about vector spaces, the axioms of vector spaces come in mighty handy. If we are proving theorems in a real analysis course, we need to have axioms about the structure of the real numbers. These additional axioms are sometimes explicitly stated and sometimes they are blanket assumptions that are made without being stated, but they are almost always there. In this section I give a couple of examples of sets of nonlogical axioms that we might use in writing deductions.

Example 2.8.1. For many of us, the first explicit set of nonlogical axioms that we see is in a course on linear algebra. To work those axioms out explicitly, let us fix the language $\mathcal{L}$ as consisting of one binary function symbol, $\oplus$, and infinitely many unary function symbols, $c\cdot$, one for each real number c. (Yes, that symbol is "c-dot.") These function symbols will be used to represent the functions of scalar multiplication. We will also have one constant symbol, 0, to represent the zero vector of the vector space. Here, then, is one way to list the nonlogical axioms of a vector space:

1. $(\forall x)(\forall y)x \oplus y = y \oplus x$ (vector addition is commutative).
2. $(\forall x)(\forall y)(\forall z)x \oplus (y \oplus z) = (x \oplus y) \oplus z$ (vector addition is associative).
3. $(\forall x)x \oplus 0 = x$.
4. $(\forall x)(\exists y)x \oplus y = 0$.
5. $(\forall x)1 \cdot x = x$.
6. $(\forall x)(c_1c_2) \cdot x = c_1 \cdot (c_2 \cdot x)$.
7. $(\forall x)(\forall y)c \cdot (x \oplus y) = c \cdot x \oplus c \cdot y$.
8. $(\forall x)(c_1 + c_2) \cdot x = c_1 \cdot x \oplus c_2 \cdot x$.

Notice a couple of things here: There are infinitely many axioms listed, as the last three axioms are really axiom schemas, consisting of one axiom for each choice of c, c_1, and c_2. An axiom schema is a template, saying that a formula is in the axiom set if it is of a certain form. Also notice that I've cheated in using the addition sign to stand for addition and juxtaposition to stand for multiplication of real numbers since the language $\mathcal{L}$ does not allow that sort of thing. See Exercise 1.

Example 2.8.2. We will write out the axioms for a dense linear order without endpoints. Our language consists of a single binary relation symbol, $<$. Our nonlogical axioms are:

1. $(\forall x)(\forall y)(x < y \vee x = y \vee y < x)$.
2. $(\forall x)(\forall y)[x = y \rightarrow \neg x < y]$.
3. $(\forall x)(\forall y)(\forall z)[(x < y \wedge y < z) \rightarrow x < z]$.
4. $(\forall x)(\forall y)\big[x < y \rightarrow \big((\exists z)(x < z \wedge z < y)\big)\big]$.
5. $(\forall x)(\exists y)(\exists z)(y < x \wedge x < z)$.

The first three axioms guarantee that the relation denoted by $<$ is a linear order, the fourth axiom states that the relation is dense, and the final axiom ensures that there is no smallest element and no greatest element.

Notice that in both of our examples, the axiom set involved is decidable: Given a formula ϕ that is alleged to be either an axiom for vector spaces or an axiom for dense linear orders without endpoints, we could decide whether or not the formula was, in fact, such an axiom. And furthermore, we could write a computer program that could decide the issue for us.

Example 2.8.3. It is time to introduce a collection of nonlogical axioms that will be vitally important to us for the rest of the book. We work in the language of number theory,

$$\mathcal{L}_{NT} = \{0, S, +, \cdot, E, <\}.$$

The set of axioms we will call N is a minimal set of assumptions to describe a bare-bones version of the usual operations on the set of natural numbers. Just how weak these axioms are will be discussed in the next chapter. These axioms will, however, be important to us in Chapters 4 and 5 precisely because they are so weak.

The Axioms of N

1. $(\forall x)\neg Sx = 0$.
2. $(\forall x)(\forall y)\left[Sx = Sy \rightarrow x = y\right]$.
3. $(\forall x)x + 0 = x$.
4. $(\forall x)(\forall y)x + Sy = S(x + y)$.
5. $(\forall x)x \cdot 0 = 0$.
6. $(\forall x)(\forall y)x \cdot Sy = (x \cdot y) + x$.
7. $(\forall x)xE0 = S0$.
8. $(\forall x)(\forall y)xE(Sy) = (xEy) \cdot x$.
9. $(\forall x)\neg x < 0$.
10. $(\forall x)(\forall y)\left[x < Sy \leftrightarrow (x < y \lor x = y)\right]$.
11. $(\forall x)(\forall y)\left[(x < y) \lor (x = y) \lor (y < x)\right]$.

Although I have just claimed that N is a weak set of axioms, let us show that N is strong enough to prove some of the basic facts about the relations and functions on the natural numbers. For the following discussion, if a is a natural number, let $\overline{a}$ be the $\mathcal{L}_{NT}$-term $\underbrace{SSS\cdots S}_{aS\text{'s}}0$. So $\overline{a}$ is the canonical term of the language that is intended to refer to the natural number a.

Lemma 2.8.4. *For natural numbers a and b:*

1. *If $a = b$, then $N \vdash \overline{a} = \overline{b}$.*
2. *If $a \neq b$, then $N \vdash \overline{a} \neq \overline{b}$.*
3. *If $a < b$, then $N \vdash \overline{a} < \overline{b}$.*
4. *If $a \not< b$, then $N \vdash \overline{a} \not< \overline{b}$.*
5. $N \vdash \overline{a} + \overline{b} = \overline{a + b}$
6. $N \vdash \overline{a} \cdot \overline{b} = \overline{a \cdot b}$
7. $N \vdash \overline{a}E\overline{b} = \overline{a^b}$

Proof: Let us begin with (1), and let us work rather carefully. Notice that the theorem is saying that if the *number* a is equal to the *number* b, then there is a deduction from the axioms in N of the formula

$$\underbrace{SS\cdots S}_{aS\text{'s}}0 = \underbrace{SS\cdots S}_{bS\text{'s}}0.$$

We work by induction on a (and b, since $a = b$). So, first assume that $a = b = 0$. Here is the needed deduction in N:

$$\begin{array}{ll} \vdots & \text{Deduction of } (\forall x)x = x \text{ (see Lemma 2.7.2)} \\ (\forall x)x = x & \\ (\forall x)x = x \to 0 = 0 & \text{(Q1)} \\ 0 = 0. & \text{(PC)} \end{array}$$

Now, what if $a = b$ and a and b are greater than 0? Then certainly $a - 1$ and $b - 1$ are equal, and by the inductive hypothesis there is a deduction of $\underbrace{SS\cdots S}_{a-1S\text{'s}}0 = \underbrace{SS\cdots S}_{b-1S\text{'s}}0$. If we follow that deduction with a use of axiom (E2): $x = y \to Sx = Sy$, and then (PC) gives us $\underbrace{SS\cdots S}_{aS\text{'s}}0 = \underbrace{SS\cdots S}_{bS\text{'s}}0$, as needed. Write out the details of the end of this deduction. It is a little trickier than I have made it sound when you actually have to use (Q1) to do the substitution. This finishes the inductive step of the proof, so (1) is established.

Looking at (2), suppose that $a \neq b$. If one of a or b is 0, then $\neg\overline{a} = \overline{b}$ follows quickly from Axiom N1 and the fact that N proves that $=$ is an equivalence relation. If neither a nor b is 0, we proceed by induction on the largest of a, b. Since $a - 1 \neq b - 1$, by the inductive hypothesis, $N \vdash \neg\overline{a-1} = \overline{b-1}$. Then by Axiom N2, $N \vdash \neg S(\overline{a-1}) = S(\overline{b-1})$. In other words, $N \vdash \neg\overline{a} = \overline{b}$, as $S(\overline{a-1})$ is typographically equivalent to $\overline{a}$ and $S(\overline{b-1})$ is typographically equivalent to $\overline{b}$.

For (3), we use induction on b. As $a < b$, we know that $b \neq 0$ and we know that $a < b - 1$ or $a = b - 1$. So either

$N \vdash \overline{a} < \overline{b-1}$ (by the inductive hypothesis)

or

$N \vdash \overline{a} = \overline{b-1}$ (by (1)).

So

$$N \vdash (\overline{a} < \overline{b-1} \lor \overline{a} = \overline{b-1}).$$

But then by Axiom N10, $N \vdash \overline{a} < S(\overline{b-1})$, which is exactly the same as $N \vdash \overline{a} < \overline{b}$.

We will now discuss (5), leaving (4), (6), and (7) to the exercises. We prove (5) by induction on b. If $b = 0$, then $\overline{a+b} = \overline{a+0} = \overline{a}$. So Axiom N3 tells us that $N \vdash \overline{a} + \overline{b} = \overline{a}$.

For the inductive step, if $b = c + 1$, then $\overline{a} + \overline{b} = \overline{a} + S(\overline{c})$. So Axiom N4 tells us that

$$N \vdash \overline{a} + \overline{b} = S(\overline{a} + \overline{c}).$$

Since $N \vdash \overline{a} + \overline{c} = \overline{a+c}$ by the inductive hypothesis, the equality axioms tell us that $N \vdash S(\overline{a} + \overline{c}) = S(\overline{a+c})$. But $S(\overline{a+c})$ *is* $\overline{a+c+1}$, which is $\overline{a+b}$. Since we know (by Theorem 2.7.1) that $N \vdash$ "equality is transitive," $N \vdash \overline{a} + \overline{b} = \overline{a+b}$. ■

2.8.1 Exercises

1. This problem is in the setting of Example 2.8.1. Exactly one of the following two statements is in the collection of nonlogical axioms of that example. Figure out which one it is, and why.

- $(\forall x)(17 + 42) \cdot x = 17 \cdot x \oplus 42 \cdot x.$
- $(\forall x)59 \cdot x = 17 \cdot x \oplus 42 \cdot x.$

Now fix up the presentation of the axioms for a vector space. You may need to redefine the language, or you may be able to take what is presented in Example 2.8.1 and fix it up.

2. For each of the following structures, decide whether or not it satisfies all of the axioms of Example 2.8.2. If the structure is *not* a dense linear order without endpoints, point out which of the axioms the structure fails to satisfy.

(a) The structure $\langle \mathbb{N}, < \rangle$, the natural numbers with the usual less than relation

(b) The structure $\langle \mathbb{Z}, < \rangle$, the integers with the usual less than relation

(c) The structure $\langle \mathbb{Q}, < \rangle$, the set of rational numbers with the usual less than relation

(d) The structure $\langle \mathbb{R}, < \rangle$, the real numbers with the usual less than relation

(e) The structure $\langle \mathbb{C}, < \rangle$, the complex numbers with the relation $<$ defined by:

$$a + bi < c + di \text{ if and only if } (a^2 + b^2) < (c^2 + d^2).$$

3. Write out the axioms for group theory. If you do not know the axioms of group theory, go to the library and check out any book with the phrase "abstract algebra", "modern algebra", or "group theory" in the title. Then check the index under "group." Specify your language carefully and then writing out the axioms should be easy.

4. In this exercise you are asked to write up some of the axioms of Zermelo–Fraenkel set theory, also known as ZF. The language of set theory consists of a single binary relation symbol, $\in$, that is intended to represent the relation "is an element of." So the formula $x \in y$ will usually be interpreted as meaning that the set x is an element of the set y. Here are English versions of some of the axioms of ZF. Write them up formally as sentences in the language of set theory.

The Axiom of Extensionality: Two sets are equal if and only if they have the same elements.

The Null Set Axiom: There is a set with no elements.

The Pair Set Axiom: If a and b are sets, then there is a set whose only elements are a and b.

The Axiom of Union: If a is a set, then there is a set consisting of exactly the elements of the elements of a. [*Query:* Can you figure out why this is called the axiom of union? Write up an example, where a is a set of three sets and each of those three sets has two elements. What does the set whose existence is guaranteed by this axiom look like?]

The Power Set Axiom: If a is a set, then there is a set consisting of all of the subsets of a. [*Suggestion:* For this axiom it might be nice to define $\subseteq$ by saying that $x \subseteq y$ is shorthand for (some nice formula with x and y free in the language of set theory).]

5. Complete the proof of Lemma 2.8.4.

6. Lemma 2.8.4(2) states that there is a deduction in N of the sentence $\neg(= S0SS0)$. Find a deduction in N of this sentence.

7. This problem is just to give you a hint of how little we can prove using the axiom system N. Suppose that we wanted to prove that $N \nvdash \neg x < x$. It makes sense (and is a consequence of the Completeness Theorem, Theorem 3.2.2) that one way to go about this would be to construct an $\mathcal{L}_{NT}$-structure $\mathfrak{A}$ in which all the axioms of N are true but $(\forall x)\neg x < x$ is not true. Do so. I would suggest that you take as your universe the set

$$A = \{0, 1, 2, 3, \dots\} \cup \{a\},$$

where a is the letter a and not a natural number. You need to define the functions $S^{\mathfrak{A}}, +^{\mathfrak{A}}$, etc., and the relation $<^{\mathfrak{A}}$. Don't do anything too strange for the natural numbers, but make sure that $a <^{\mathfrak{A}} a$. Check that the axioms of N are true in the structure $\mathfrak{A}$, and you're finished!

8. Using more or less the same technique as in Exercise 7, show that N does not prove that addition is commutative.

2.9 Summing Up, Looking Ahead

In these first two chapters we have developed a vocabulary for talking about mathematical structures, mathematical languages, and deductions. Chapter 2 has focused on deductions, which are supposed to be the formal equivalents of the mathematical proofs that you have seen for many years. We have seen some results, such as the Deduction Theorem, which indicate that deductions behave like proofs behave. The Soundness Theorem shows that deductions preserve truth, which gives us some comfort as we try to justify in our minds why proofs preserve truth.

As you look at the statement of the Soundness Theorem, you can see that it is explicitly trying to relate the syntactical notion of deducibility ($\vdash$) with the semantical notion of logical implication

($\models$). The first major result of Chapter 3, the Completeness Theorem, will also relate these two notions and will in fact show that they are equivalent. Then the Compactness Theorem (which is really a quite trivial consequence of the Completeness Theorem) will be used to construct some mathematical models with some very interesting properties.

Chapter 3

Completeness and Compactness

3.1 Naïvely

We are at a point in our explorations where we have established a particular deductive system, consisting of the logical axioms and rules of inference that we set out in the last chapter. The Soundness Theorem showed that our deductive system preserves truth, in the sense that if there is a deduction of ϕ from Σ, then ϕ is true in any model of Σ. The Completeness Theorem, the first major result of this chapter, gives us the converse to the Soundness Theorem. So, when the two results are combined, we will have this equivalence:

$$\Sigma \models \phi \text{ if and only if } \Sigma \vdash \phi.$$

We have already made a big point of the fact that we would like to be sure that if our deductive system allows us to prove a statement, we would like that statement to be true. Certainly, the content of the Soundness Theorem is exactly that. If $\vdash \phi$, if there is a deduction of ϕ from only the logical axioms without any additional assumptions, then we know that $\models \phi$, so ϕ is true in every structure with every assignment function. To the extent that the informal mathematical practice of everyday proofs is modeled by our formal system of deduction, we can be sure that the things that we prove mathematically are true.

If life were peaches and cream, we would also like to know that we can prove anything that is true. The Completeness Theorem is the result that asserts that our deductive system *is* that strong. So you would be tempted to conclude that, for example, we are able to prove any statement of first-order logic that is a true statement about the natural numbers.

Unfortunately, this conclusion is based upon a misreading of the statement of the Completeness Theorem. What we will prove is that our deductive system is complete, in the sense of this definition:

Definition 3.1.1. A deductive system consisting of a collection of logical axioms Λ and a collection of rules of inference is said to be **complete** if for every set of nonlogical axioms Σ and every $\mathcal{L}$-formula ϕ,

$$\text{If } \Sigma \models \phi, \text{ then } \Sigma \vdash \phi.$$

What this says is that if ϕ is an $\mathcal{L}$-formula that is true in *every* model of Σ, then there will be a deduction from Σ of ϕ. So our ability to prove ϕ depends on ϕ being true in every model of Σ. Thus if we want to be able to use Σ to prove every true statement about the natural numbers, we have to be able to find a set of nonlogical axioms Σ such that $\Sigma \models \phi$ if and only if ϕ is a true statement about the natural numbers. We will have much more to say about that problem in Chapters 4 and 5.

The second part of the chapter concerns the Compactness Theorem and the Löwenheim–Skolem Theorems. We will use these results to investigate various types of mathematical structures, including structures that are quite surprising.

In some sense, we have spent a lot of time in the first couple of chapters of this book developing a lot of vocabulary and establishing some basic results. Now we will roll up our sleeves and get a couple of worthwhile theorems. It is time to start showing some of the beauty and the power, as well as the limitations, of first-order logic.

3.2 Completeness

Let us fix a collection of nonlogical axioms, Σ. Our goal in this section is to show that for any formula ϕ, if $\Sigma \models \phi$, then $\Sigma \vdash \phi$. In some sense, this is the only possible interpretation of the phrase "you can prove anything that is true," if you are discussing the adequacy of the deductive system. To say that ϕ is true whenever Σ is a collection of true axioms is precisely to say that Σ logically implies ϕ. Thus, the Completeness Theorem will say that whenever ϕ is logically implied by Σ, there is a deduction from Σ of ϕ. So the Completeness Theorem is the converse of the Soundness Theorem.

We have to begin with a short discussion of consistency.

Definition 3.2.1. Let Σ be a set of $\mathcal{L}$-formulas. We will say that Σ is **inconsistent** if there is a deduction from Σ of $[(\forall x)x = x] \wedge \neg[(\forall x)x = x]$. We say that Σ is **consistent** if it is not inconsistent.

So Σ is inconsistent if Σ proves a contradiction. Exercise 1 asks you to show that if Σ is inconsistent, then there is deduction from Σ of every $\mathcal{L}$-formula. For notational convenience, let us agree to use the symbol $\perp$ (read "false" or "eet") for the contradictory sentence $[(\forall x)x = x] \wedge \neg[(\forall x)x = x]$. All you will have to remember is that $\perp$ is a sentence that is in every language and is true in no structure.

Theorem 3.2.2 (Completeness Theorem). *Suppose that Σ is a set of $\mathcal{L}$-formulas and ϕ is an $\mathcal{L}$-formula. If $\Sigma \models \phi$, then $\Sigma \vdash \phi$.*

Proof:

Chaff: This theorem was established in 1929 by the Austrian mathematician Kurt Gödel, in his Ph.D. dissertation. If you haven't picked it up already, you should know that the work of Gödel is central to the development of logic in the twentieth century. He is responsible for most of the major results that we will state in the rest of the book: The Completeness Theorem, the Compactness Theorem, and the two Incompleteness Theorems. Gödel was an absolutely brilliant man, with a complex and troubled personality. A wonderful and engaging biography of Gödel is [Dawson 97]. The first volume of Gödel's collected works, [Gödel–Works], also includes a biography and introductory comments about his papers that can help your understanding of this wonderful mathematics.

The proof we present of the Completeness Theorem is based on work of Leon Henkin. The idea of Henkin's proof is brilliant, but the details take some time to work through. Just to warn you, this proof doesn't end until page 99.

Before we get involved in the details, let us look at a rough outline of how the argument proceeds. There are a few simplifications and one or two outright lies in the outline, but we will straighten everything out as we work out the proof.

Outline of the Proof

There will be a **preliminary argument** that will show that it is sufficient to prove that if Σ is a consistent set of sentences, then Σ has a model. Then we will proceed to assume that we are given such a set of sentences, and we will construct a model for Σ.

The construction of the model will proceed in several steps, but the central idea was introduced in Example 1.6.4. The elements of the model will be variable-free terms of a language. We will construct this model so that the formulas that will be true in the model are precisely the formulas that are in a certain set of formulas, which we will call Σ'. We will make sure that $\Sigma \subseteq \Sigma'$, so all of the formulas of Σ will be true in this constructed model. In other words, we will have constructed a model of Σ.

To make the construction work we will take our given set of $\mathcal{L}$-sentences Σ and extend it to a bigger set of sentences Σ' in a bigger language $\mathcal{L}'$. We do this extension in two steps. First, we will add in some new axioms, called Henkin Axioms, to get a collection $\hat{\Sigma}$. Then we will extend $\hat{\Sigma}$ to Σ' in such a way that:

1. Σ' is consistent.

2. For every $\mathcal{L}'$-sentence θ, either $\theta \in \Sigma'$ or $(\neg\phi) \in \Sigma'$.

Thus we will say that Σ' is a maximal consistent extension of Σ, where *maximal* means that it is impossible to add any sentences to Σ' without making Σ' inconsistent.

Now there are two possible sources of problems in this expansion of Σ to Σ'. The first is that we will **change languages from $\mathcal{L}$ to $\mathcal{L}'$**, where $\mathcal{L} \subseteq \mathcal{L}'$. It is conceivable that Σ will not be consistent when viewed as a set of $\mathcal{L}'$-sentences, even though Σ is consistent when viewed as a set of $\mathcal{L}$-sentences. The reason that this might happen is that there are more $\mathcal{L}'$-deductions than there are $\mathcal{L}$-deductions, and one of these new deductions just might happen to be a deduction of $\bot$. Fortunately, Lemma 3.2.3 will show us that this does not happen, so Σ is consistent as a set of $\mathcal{L}'$-sentences. The other possible problem is in our two **extensions of** Σ, first to $\hat{\Sigma}$ and then to Σ'. It certainly might happen that we could add a sentence to Σ in such a way as to make Σ' inconsistent. But Lemma 3.2.4 and Exercise 4 will prove that Σ' is still consistent.

Once we have our maximal consistent set of sentences Σ', we will **construct a model** $\mathfrak{A}$ and prove that the sentences of $\mathcal{L}'$ that are in Σ' are precisely the sentences that are true in $\mathfrak{A}$. Thus, $\mathfrak{A}$ will be a model of Σ', and as $\Sigma \subseteq \Sigma'$, $\mathfrak{A}$ will be a model of Σ, as well.

This looks daunting, but if we keep our wits about us and do things one step at a time, it will all come together at the end.

Preliminary Argument

So let us fix our setting for the rest of this proof. We are working in a language $\mathcal{L}$. For the purposes of this proof, we assume that the language is countable, which means that the formulas of $\mathcal{L}$ can be written in an infinite list $\alpha_1, \alpha_2, \ldots, \alpha_n, \ldots$. (An outline of the changes in the proof necessary for the case when $\mathcal{L}$ is not countable can be found in Exercise 6.)

We are given a set of formulas Σ, and we are assuming that $\Sigma \models \phi$. We have to prove that $\Sigma \vdash \phi$.

Note that we can assume that ϕ is a sentence: By Lemma 2.7.2, $\Sigma \vdash \phi$ if and only if there is a deduction from Σ of the universal closure of ϕ. Also, by the comments following Lemma 2.7.3, we can also assume that every element of Σ is a sentence. So, now all(!) we have to do is prove that if Σ is a set of *sentences* and ϕ is a *sentence* and if $\Sigma \models \phi$, then $\Sigma \vdash \phi$.

Now we claim that it suffices to prove the case where ϕ is the sentence $\bot$. For suppose we know that if $\Sigma \models \bot$, then $\Sigma \vdash \bot$, and suppose we are given a sentence ϕ such that $\Sigma \models \phi$. Then $\Sigma \cup (\neg\phi) \models \bot$, as there are no models of $\Sigma \cup (\neg\phi)$, so $\Sigma \cup (\neg\phi) \vdash \bot$. This tells us, by Exercise 4 in Section 2.7.1, that $\Sigma \vdash \phi$, as needed.

So we have reduced what we need to do to proving that if $\Sigma \models \bot$, then $\Sigma \vdash \bot$, for Σ a set of $\mathcal{L}$-sentences. This is equivalent to saying that if there is no model of Σ, then $\Sigma \vdash \bot$. We will work with the contrapositive: If $\Sigma \nvdash \bot$, then there is a model of Σ. In other words, we will prove:

If Σ is a consistent set of sentences, then there is a model of Σ.

This ends the preliminary argument that was promised in the outline of the proof. Now, we will assume that Σ is a consistent set of $\mathcal{L}$-sentences and go about the task of constructing a model of Σ.

Changing the Language from $\mathcal{L}$ to $\mathcal{L}_1$

The model of Σ that we will construct will be a model whose elements are variable-free terms of a language. This might lead to problems. For example, suppose that $\mathcal{L}$ contains no constant symbols. Then there will be no variable-free terms of $\mathcal{L}$. Or, perhaps $\mathcal{L}$ has exactly one constant symbol c, no function symbols, one unary relation P, and

$$\Sigma = \{\exists x P(x), \neg P(c)\}.$$

Here Σ is consistent, but no structure whose universe is $\{c\}$ (c is the only variable-free term of $\mathcal{L}$) can be a model of Σ. So we have to expand our language to give us enough constant symbols to build our model.

So let $\mathcal{L}_0 = \mathcal{L}$, and define

$$\mathcal{L}_1 = \mathcal{L}_0 \cup \{c_1, c_2, \ldots, c_n, \ldots\},$$

where the c_i's are new constant symbols.

> *Chaff:* Did you notice that when I was defining $\mathcal{L}_1$ I took something we already knew about, $\mathcal{L}$, and gave it a new name, $\mathcal{L}_0$? When you are reading mathematics and something like that happens, it is almost always a clue that whatever happens next is going to be iterated, in this case to build $\mathcal{L}_2, \mathcal{L}_3$, and so on. Back in my English lit course we called that foreshadowing.

We say (for the obvious reason) that $\mathcal{L}_1$ is an **extension by constants** of $\mathcal{L}_0$. As mentioned in the outline, it is not immediately clear that Σ remains consistent when viewed as a collection of $\mathcal{L}_1$-sentences rather than $\mathcal{L}$-sentences. The following lemma, the proof of which is delayed until page 99, shows that Σ remains consistent.

Lemma 3.2.3. *If Σ is a consistent set of $\mathcal{L}$-sentences and $\mathcal{L}_1$ is an extension by constants of $\mathcal{L}$, then Σ is consistent when viewed as a set of $\mathcal{L}_1$-sentences.*

The constants that we have added to form $\mathcal{L}_1$ are called **Henkin constants**, and they serve a particular purpose. They will be the witnesses that allow us to ensure that any time Σ claims $\exists x \phi(x)$, then in our constructed model $\mathfrak{A}$, there will be an element (which will be one of these constants c) such that $\mathfrak{A} \models \phi(c)$.

> *Chaff:* Recall that the notation $\exists x \phi(x)$ implies that ϕ is a formula with x as the only free variable. Then $\phi(c)$ is the result of replacing the free occurrences of x with the constant symbol c. Thus $\phi(c)$ is ϕ^x_c.

The next step in our construction makes sure that the Henkin constants will be the witnesses for the existential sentences in Σ.

Extending Σ to Include Henkin Axioms

Consider the collection of sentences of the form $\exists x \theta$ in the language $\mathcal{L}_0$. As the language $\mathcal{L}_0$ is countable, the collection of $\mathcal{L}_0$-sentences is countable, so we can list all such sentences of the form $\exists x \theta$, enumerating them by the positive integers:

$$\exists x \theta_1, \exists x \theta_2, \exists x \theta_3, \ldots, \exists x \theta_n, \ldots.$$

We will now use the Henkin constants of $\mathcal{L}_1$ to add to Σ countably many axioms, called **Henkin axioms**. These axioms will ensure that every existential sentence that is asserted by Σ will have a witness in our constructed structure $\mathfrak{A}$. The collection of Henkin axioms is

$$H_1 = \{[\exists x \theta_i] \rightarrow \theta_i(c_i) \mid (\exists x \theta_i) \text{ is an } \mathcal{L}_0 \text{ sentence}\},$$

where $\theta_i(c_i)$ is shorthand for $\theta^x_{c_i}$.

Now let $\Sigma_0 = \Sigma$, and define

$$\Sigma_1 = \Sigma_0 \cup H_1.$$

> *Chaff:* Foreshadowing!

As Σ_1 contains many more sentences than Σ_0, it seems entirely possible that Σ_1 is no longer consistent. Fortunately, the next lemma shows that is not the case. The proof of the lemma is on page 100.

Lemma 3.2.4. *If Σ_0 is a consistent set of sentences and Σ_1 is created by adding Henkin axioms to Σ_0, then Σ_1 is consistent.*

Now we have Σ_1, a consistent set of $\mathcal{L}_1$-sentences. We can repeat this construction, building a larger language $\mathcal{L}_2$ consisting of $\mathcal{L}_1$ together with an infinite set of new Henkin constants k_i. Then we can let H_2 be a new set of Henkin axioms:

$$H_2 = \{[\exists x \theta_i] \rightarrow \theta_i(k_i) \mid (\exists x \theta_i) \text{ is an } \mathcal{L}_1 \text{ sentence}\},$$

and let Σ_2 be $\Sigma_1 \cup H_2$. As before, Σ_2 will be consistent. We can continue this process to build:

- $\mathcal{L} = \mathcal{L}_0 \subseteq \mathcal{L}_1 \subseteq \mathcal{L}_2 \cdots$, an increasing chain of languages.
- $H_1, H_2, H_3, \ldots$, each H_i a collection of Henkin axioms in the language $\mathcal{L}_i$.
- $\Sigma = \Sigma_0 \subseteq \Sigma_1 \subseteq \Sigma_2 \subseteq \cdots$, where each Σ_i is a consistent set of $\mathcal{L}_i$-sentences.

Let $\mathcal{L}' = \bigcup_{i<\infty} \mathcal{L}_i$ and let $\hat{\Sigma} = \bigcup_{i<\infty} \Sigma_i$. Each Σ_i is a consistent set of $\mathcal{L}'$-sentences, as can be shown by proofs that are identical to those of Lemmas 3.2.3 and 3.2.4. You will show in Exercise 2 that $\hat{\Sigma}$ is a consistent set of $\mathcal{L}'$-sentences.

Extending to a Maximal Consistent Set of Sentences

As you recall, we were going to construct our model $\mathfrak{A}$ in such a way that the sentences that were true in $\mathfrak{A}$ were exactly the elements of a set of sentences Σ'. It is time to build Σ'. Since every sentence is either true or false in a given model, it will be necessary for us to make sure that for every sentence $\sigma \in \mathcal{L}'$, either $\sigma \in \Sigma'$ or $\neg\sigma \in \Sigma'$. Since we can't have both σ and $\neg\sigma$ true in any structure, we must also make sure that we don't put both σ and $\neg\sigma$ into Σ'. Thus, Σ' will be a maximal consistent extension of $\hat{\Sigma}$.

To build this extension, fix an enumeration of all of the $\mathcal{L}'$-sentences

$$\sigma_1, \sigma_2, \ldots, \sigma_n, \ldots.$$

We can do this as $\mathcal{L}'$ is countable, being a countable union of countable sets. Now we work our way through this list, one sentence at a time, adding either σ_n or the denial of σ_n to our growing list of sentences, depending on which one keeps our collection consistent.

Here are the details. Let $\Sigma^0 = \hat{\Sigma}$, and assume that Σ^k is known to be a consistent set of $\mathcal{L}'$-sentences. We will show how to build $\Sigma^{k+1} \supseteq \Sigma^k$ and prove that Σ^{k+1} is also a consistent set of $\mathcal{L}'$-sentences. Then we let

$$\Sigma' = \Sigma^0 \cup \Sigma^1 \cup \Sigma^2 \cup \cdots \cup \Sigma^n \cup \cdots .$$

You will prove in Exercise 4 that Σ' is a consistent set of sentences. It will be obvious from the construction of Σ^{k+1} from Σ^k that Σ' is maximal, and thus we will have completed our task of producing a maximal consistent extension of $\hat{\Sigma}$.

So all we have to do is describe how to get Σ^{k+1} from Σ^k and prove that Σ^{k+1} is consistent. Given Σ^k, consider the set $\Sigma^k \cup \{\sigma^{k+1}\}$, where σ_{k+1} is the $(k+1)$st element of our fixed list of all of the $\mathcal{L}'$-sentences. Let

$$\Sigma^{k+1} = \begin{cases} \Sigma^k \cup \{\sigma_{k+1}\} & \text{if } \Sigma^k \cup \{\sigma_{k+1}\} \text{ is consistent,} \\ \Sigma^k \cup \{\neg\sigma_{k+1}\} & \text{otherwise.} \end{cases}$$

You are asked in Exercise 3 to prove that Σ^{k+1} is consistent. Once you have done that, we have constructed a maximal consistent Σ' that extends Σ.

The next lemma states that Σ' is deductively closed, at least as far as sentences are concerned. As you work through the proof, the Deduction Theorem will be useful.

Lemma 3.2.5. *If σ is a sentence, then $\sigma \in \Sigma'$ if and only if $\Sigma' \vdash \sigma$.*

Proof: Exercise 5. ■

Construction of the Model—Preliminaries

I have mentioned a few times that the model of Σ that we are going to construct will have as its universe the collection of variable-free terms of the language $\mathcal{L}'$. It is now time to confess that I have lied. It is easy to see why the plan of using the terms as the elements of the universe is doomed to failure. Suppose that there are two different terms t_1 and t_2 of the language and somewhere in Σ' is the sentence $t_1 = t_2$. If the *terms* were the elements of the universe, then we could not model Σ', as the two terms t_1 and t_2 are not the same (they are typographically distinct), while Σ' demands that they be equal. Our solution to this problem is to take the collection of variable-free terms, define an equivalence relation on that set, and then construct a model from the *equivalence classes* of the variable-free terms.

So let T be the set of variable-free terms of the language $\mathcal{L}'$, and define a relation $\sim$ on T by

$$t_1 \sim t_2 \text{ if and only if } (t_1 = t_2) \in \Sigma'.$$

It is not difficult to show that $\sim$ is an equivalence relation. We will verify that $\sim$ is symmetric, leaving reflexivity and transitivity to the Exercises.

To show that $\sim$ is symmetric, assume that $t_1 \sim t_2$. We must prove that $t_2 \sim t_1$. As we know $t_1 \sim t_2$, by definition we know that the sentence $(t_1 = t_2)$ is an element of Σ'. We need to show that $(t_2 = t_1) \in \Sigma'$. Assume not. Then by the maximality of Σ', $\neg(t_2 = t_1) \in \Sigma'$. But since we know that $\Sigma' \vdash t_1 = t_2$, by Theorem 2.7.1, $\Sigma' \vdash t_2 = t_1$. (Can you provide the details?) But since we also know that $\Sigma' \vdash \neg(t_2 = t_1)$, it must be the case that $\Sigma' \vdash \perp$, which is a contradiction, as we know that Σ' is consistent. So our assumption is wrong and $(t_2 = t_1) \in \Sigma'$, and thus $\sim$ is a symmetric relation.

So, assuming that you have worked through Exercise 7, we have established that $\sim$ is an equivalence relation. Now let $[t]$ be the set of all variable-free terms s of the language $\mathcal{L}'$ such that $t \sim s$. So $[t]$ is the equivalence class of all terms that Σ' tells us are equal to t. The collection of all such equivalence classes will be the universe of our model $\mathfrak{A}$.

Construction of the Model—The Main Ideas

To define our model of Σ', we must construct an $\mathcal{L}'$-structure. Thus, we have to describe the universe of our structure as well as interpretations of all of the constant, function, and relation symbols of the language $\mathcal{L}'$. We discuss each of them separately.

The Universe A: As explained above, the universe of $\mathfrak{A}$ will be the collection of $\sim$-equivalence classes of the variable-free terms of $\mathcal{L}'$. For example, if $\mathcal{L}'$ includes the binary function symbol f, the non-Henkin constant symbol k, and the Henkin constants $c_1, c_2, \ldots, c_n, \ldots$, then the universe of our structure would include among its elements $[c_{17}]$ and $[f(k, c_3)]$.

The Constants: For each constant symbol c of $\mathcal{L}'$ (including the Henkin constants), we need to pick out an element $c^{\mathfrak{A}}$ of the universe to be the element represented by that symbol. We don't do anything fancy here:

$$c^{\mathfrak{A}} = [c].$$

So each constant symbol will denote its own equivalence class.

The Functions: If f is an n-ary function symbol, we must define an n-ary function $f^{\mathfrak{A}} : A^n \to A$. Let me write down the definition of $f^{\mathfrak{A}}$ and then we can try to figure out exactly what the definition is saying:

$$f^{\mathfrak{A}}([t_1], [t_2], \ldots, [t_n]) = [ft_1t_2 \ldots t_n].$$

On the left-hand side of the equality you will notice that there are n equivalence classes that are the inputs to the function $f^{\mathfrak{A}}$. Since the elements of A are equivalence classes and $f^{\mathfrak{A}}$ is an n-ary function, that should be all right. On the right side of the equation there is a single equivalence class, and the thing inside the brackets is a variable-free term of $\mathcal{L}'$. Notice that the function $f^{\mathfrak{A}}$ acts by placing the symbol f in front of the terms and then taking the equivalence class of the result.

There is one detail that has to be addressed. We must show that the function $f^{\mathfrak{A}}$ is well defined. Let me say a bit about what that means, assuming that f is a unary function symbol, for simplicity. Notice that our definition of $f^{\mathfrak{A}}([t])$ depends on the *name* of the equivalence class that we are putting into $f^{\mathfrak{A}}$. This might lead to problems, as it is at least conceivable that we could have two terms, t_1 and t_2, such that $[t_1]$ is the same set as $[t_2]$, but $f^{\mathfrak{A}}([t_1])$ and $f^{\mathfrak{A}}([t_2])$ evaluate to be different sets. Then our alleged function $f^{\mathfrak{A}}$ wouldn't even be a function. Showing that this does not happen is what we mean when we say that we must show that the function $f^{\mathfrak{A}}$ is well defined.

Let us look at the proof that our function $f^{\mathfrak{A}}$ is, in fact, well defined. Suppose that $[t_1] = [t_2]$. We must show that $f^{\mathfrak{A}}([t_1]) = f^{\mathfrak{A}}([t_2])$. In other words, we must show that if $[t_1] = [t_2]$, then $[ft_1] = [ft_2]$. Again looking at the definition of our equivalence relation $\sim$, this means that we must show that if $t_1 = t_2$ is an element of Σ', then so is $f(t_1) = f(t_2)$. So assume that $t_1 = t_2$ is an element of Σ'. Here is an outline of a deduction from Σ' of $f(t_1) = f(t_2)$:

$x = y \rightarrow f(x) = f(y)$	axiom (E2)
$\vdots$	
$t_1 = t_2 \rightarrow f(t_1) = f(t_2)$	
$t_1 = t_2$	element of Σ'
$f(t_1) = f(t_2)$	PC

Since $\Sigma' \vdash f(t_1) = f(t_2)$, Lemma 3.2.5 tells us that $f(t_1) = f(t_2)$ is an element of Σ', as needed. So the function $f^{\mathfrak{A}}$ is well defined.

The Relations: Suppose that R is an n-ary relation symbol of $\mathcal{L}'$. We must define an n-ary relation $R^{\mathfrak{A}}$ on A. In other words, we must decide which n-tuples of equivalence classes will stand in the relation $R^{\mathfrak{A}}$. Here is where we use the elements of Σ'. We define $R^{\mathfrak{A}}$ by this statement:

$$R^{\mathfrak{A}}([t_1], [t_2], \ldots, [t_n]) \text{ is true if and only if } Rt_1t_2\ldots t_n \in \Sigma'.$$

So elements of the universe are in the relation R if and only if Σ' *says* they are in the relation R. Of course, we must show that the relation $R^{\mathfrak{A}}$ is well defined, also. Or rather, **you** must show that the relation $R^{\mathfrak{A}}$ is well defined. See Exercise 8.

At this point we have constructed a perfectly good $\mathcal{L}'$-structure. What we have to do next is show that $\mathfrak{A}$ makes all of the sentences of Σ' true. Then we will have shown that we have constructed a model of Σ'.

Proposition 3.2.6. $\mathfrak{A} \models \Sigma'$.

Proof: We will in fact prove something slightly stronger. We will prove, for each sentence σ, that

$$\sigma \in \Sigma' \text{ if and only if } \mathfrak{A} \models \sigma.$$

Well, since you have noticed, this isn't *really* stronger, as we know that Σ' is maximal. But it does appear stronger, and this version of the proposition is what we need to get the inductive steps to work out nicely.

We proceed by induction on the complexity of the formulas in Σ'. For the base case, suppose that σ is an atomic sentence. Then σ is of the form $Rt_1t_2 \ldots t_n$, where R is an n-ary relation symbol and the t_i's are variable free terms. But then our definition of $R^{\mathfrak{A}}$ guaranteed that $\mathfrak{A} \models \sigma$ if and only if $\sigma \in \Sigma'$. Notice that if R is $=$ and σ is $t_1 = t_2$, then $\sigma \in \Sigma'$ iff $t_1 \sim t_2$ iff $[t_1] = [t_2]$ iff $\mathfrak{A} \models \sigma$.

For the inductive cases, suppose first that σ is $\neg\alpha$, where we know by inductive hypothesis that $\mathfrak{A} \models \alpha$ if and only if $\alpha \in \Sigma'$. Notice that as Σ' is a maximal consistent set of sentences, we know that $\sigma \in \Sigma'$ if and only if $\alpha \notin \Sigma'$. Thus

$$\begin{aligned} \sigma \in \Sigma' & \text{ if and only if } \alpha \notin \Sigma' \\ & \text{ if and only if } \mathfrak{A} \not\models \alpha \\ & \text{ if and only if } \mathfrak{A} \models \neg\alpha \\ & \text{ if and only if } \mathfrak{A} \models \sigma. \end{aligned}$$

The second inductive case, when σ is $\alpha \vee \beta$, is similar and is left to the Exercises.

The interesting case is when σ is a sentence of the form $\forall x\phi$. We must show that $\forall x\phi \in \Sigma'$ if and only if $\mathfrak{A} \models \forall x\phi$. We do each implication separately.

First, assume that $\forall x\phi \in \Sigma'$. We must show that $\mathfrak{A} \models \forall x\phi$, which means that we must show, given an assignment function s, that $\mathfrak{A} \models \forall x\phi[s]$. Since the elements of A are equivalence classes of variable-free terms, this means that we have to show for any variable-free term t that

$$\mathfrak{A} \models \phi\big[s[x|[t]]\big].$$

But (here is another lemma for you to prove) for any variable-free term t and any assignment function s, $\overline{s}(t) = [t]$, and so by Theorem 2.6.2, we need to prove that

$$\mathfrak{A} \models \phi^x_t[s].$$

Notice that ϕ^x_t is a sentence, so $\mathfrak{A} \models \phi^x_t[s]$ if and only if $\mathfrak{A} \models \phi^x_t$. But also notice that $\Sigma' \vdash \phi^x_t$, as $\forall x\phi$ is an element of Σ', $\forall x\phi \rightarrow$

ϕ_t^x is a quantifier axiom of type (Q1) (t is substitutable for x in ϕ as t is variable-free), and Σ' is deductively closed for sentences by Lemma 3.2.5. But ϕ_t^x is less complex than $\forall x \phi$, and thus by our inductive hypothesis, $\mathfrak{A} \models \phi_t^x$, as needed.

For the reverse direction of our biconditional, assume that $\forall x \phi \notin \Sigma'$. We need to show that $\mathfrak{A} \not\models \forall x \phi$. As Σ' is maximal, $\neg \forall x \phi \in \Sigma'$. By deductive closure again, this means that $\exists x \neg \phi \in \Sigma'$. From our construction of Σ', we know there is some Henkin constant c_i such that $\left([\exists x \neg \phi] \rightarrow \neg \phi(c_i)\right) \in \Sigma'$, and using deductive closure once again, this tells us that $\neg \phi(c_i) \in \Sigma'$. Having stripped off a quantifier, we can assert via the inductive hypothesis that $\mathfrak{A} \models \neg \phi(c_i)$, so $\mathfrak{A} \not\models \forall x \phi$, as needed.

This finishes our proof of Lemma 3.2.6, so we know that the $\mathcal{L}'$-structure $\mathfrak{A}$ is a model of Σ'. ■

Construction of the Model—Cleaning Up

As you recall, back in our outline of the proof of the Completeness Theorem on page 88, we were going to prove the theorem by constructing a model of Σ. We are almost there. We have a structure, $\mathfrak{A}$, we know that $\mathfrak{A}$ is a model of Σ', and we know that $\Sigma \subseteq \Sigma'$, so every sentence in Σ is true in the structure $\mathfrak{A}$. We're just about done. The only problem is that Σ began life as a set of $\mathcal{L}$-sentences, while $\mathfrak{A}$ is an $\mathcal{L}'$-structure, not an $\mathcal{L}$-structure.

Fortunately, this is easily remedied by a slight bit of amnesia: Define the structure $\mathfrak{A} \restriction_{\mathcal{L}}$ (read $\mathfrak{A}$ **restricted to** $\mathcal{L}$, or the **restriction of** $\mathfrak{A}$ **to** $\mathcal{L}$ as follows: The universe of $\mathfrak{A} \restriction_{\mathcal{L}}$ is the same as the universe of $\mathfrak{A}$. Any constant symbols, function symbols, and relations symbols of $\mathcal{L}$ are interpreted in $\mathfrak{A} \restriction_{\mathcal{L}}$ exactly as they were interpreted in $\mathfrak{A}$, and we just ignore all of the symbols that were added as we moved from $\mathcal{L}$ to $\mathcal{L}'$. Now, $\mathfrak{A} \restriction_{\mathcal{L}}$ is a perfectly good $\mathcal{L}$-structure, and all that is left to finish the proof of the Completeness Theorem is to work through one last lemma:

Lemma 3.2.7. *If σ is an $\mathcal{L}$-sentence, then $\mathfrak{A} \models \sigma$ if and only if $\mathfrak{A} \restriction_{\mathcal{L}} \models \sigma$.*

Proof: (Outline) Use induction on the complexity of σ, proving that $\mathfrak{A} \restriction_{\mathcal{L}} \models \sigma$ if and only if $\sigma \in \Sigma'$, as in the proof of Lemma 3.2.6. ■

Thus, we have succeeded in producing an $\mathcal{L}$-structure that is a model of Σ, so we know that every consistent set of sentences has a model. By our preliminary remarks on page 89, we thus know that if $\Sigma \models \phi$, then $\Sigma \vdash \phi$, and our proof of the Completeness Theorem is complete. ■

Proofs of the Lemmas

We present here the proofs of two lemmas that were used in the proof of the Completeness Theorem. The first lemma was introduced when we expanded the language $\mathcal{L}$ to the language $\mathcal{L}'$ and we were concerned about the consistency of Σ in the new, expanded language.

(Lemma 3.2.3). *If Σ is a consistent set of $\mathcal{L}$-sentences and $\mathcal{L}'$ is an extension by constants of $\mathcal{L}$, then Σ is consistent when viewed as a set of $\mathcal{L}'$-sentences.*

Proof: Suppose, by way of contradiction, that Σ is not consistent as a set of $\mathcal{L}'$-sentences. Thus there is a deduction (in $\mathcal{L}'$) of $\bot$ from Σ. Let n be the smallest number of new constants used in any such deduction, and let D' be a deduction using exactly n such constants. Notice that $n > 0$, as otherwise D' would be a deduction of $\bot$ in $\mathcal{L}$. We show that there is a deduction of $\bot$ using fewer than n constants, a contradiction that establishes the lemma.

Let v be a variable that does not occur in D', let c be one of the new constants that occurs in D', and let D be the sequence of formulas $\langle \phi_i \rangle$ that is formed by taking each formula ϕ_i' in D' and replacing all occurrences of c in ϕ_i' by v. The last formula in D is $\bot$, so if we can show that D is a deduction, we will be finished.

So we use induction on the elements of the deduction D'. If ϕ_i' is an element of D' by virtue of being an equality axiom or an element of Σ, then $\phi_i = \phi_i'$, and ϕ_i is an element of a deduction by the same reason. If ϕ_i' is a quantifier axiom, for example $(\forall x)\theta' \to \theta'^{x}_{t'}$, then ϕ_i will also be a quantifier axiom, in this case $(\forall x)\theta \to \theta^x_t$. There will be no problems with substitutability of t for x, given that t' is substitutable for x. If ϕ' is an element of the deduction by virtue of being the conclusion of a rule of inference $\langle \Gamma', \phi' \rangle$, then $\langle \Gamma, \phi \rangle$ will be a rule of inference that will justify ϕ.

This completes the argument that D is a deduction of $\bot$. Since D clearly uses fewer new constant symbols than D', we have our contradiction and our proof is complete. ■

The second lemma was needed when we added the Henkin axioms to our consistent set of sentences Σ. We needed to prove that the resulting set, $\hat{\Sigma}$, was still consistent.

(Lemma 3.2.4). *If Σ is a consistent set of sentences and $\hat{\Sigma}$ is created by adding Henkin axioms to Σ, then $\hat{\Sigma}$ is consistent.*

Proof: Suppose that $\hat{\Sigma}$ is not consistent. Let n be the smallest number of Henkin axioms used in any possible deduction from $\hat{\Sigma}$ of $\bot$. Fix such a set of n Henkin axioms, and let α be one of those Henkin axioms. So we know that

$$\Sigma \cup H \cup \alpha \vdash \bot,$$

where H is the collection of the other $n-1$ Henkin axioms needed in the proof. Now α is of the form $\exists x \phi \to \phi(c)$, where c is a Henkin constant and $\phi(c)$ is our shorthand for ϕ^x_c.

By the Deduction Theorem (Theorem 2.7.4), as α is a sentence, this means that $\Sigma \cup H \vdash \neg\alpha$, so

$$\Sigma \cup H \vdash \exists x \phi \quad \text{and} \quad \Sigma \cup H \vdash \neg\phi^x_c.$$

Since $\exists x \phi$ is the same as $\neg\forall x \neg\phi$, from the first of these facts we know that

$$\Sigma \cup H \vdash \neg\forall x \neg\phi. \tag{3.1}$$

We also know that $\Sigma \cup H \vdash \neg\phi^x_c$. If we take a deduction of $\neg\phi^x_c$ and replace each occurrence of the constant c by a new variable z, the result is still a deduction (as in the proof of Lemma 3.2.3 above), so $\Sigma \cup H \vdash \neg\phi^x_z$. By Lemma 2.7.2, we know that

$$\Sigma \cup H \vdash \forall z \neg\phi^x_z.$$

Our quantifier axiom (Q1) states that as long as x is substitutable for z in $\neg\phi^x_z$, (which it is, as z is a *new* variable), then we may assert that $\left[\forall z \neg\phi^x_z\right] \to \neg(\phi^x_z)^z_x$. Therefore

$$\Sigma \cup H \vdash \neg(\phi^x_z)^z_x.$$

But $(\phi^x_z)^z_x = \phi$, so $\Sigma \cup H \vdash \neg\phi$. But now we can use Lemma 2.7.2 again to conclude that

$$\Sigma \cup H \vdash \forall x \neg \phi. \tag{3.2}$$

So, by Equations (3.1) and (3.2), we see that $\Sigma \cup H \vdash \bot$. This is a contradiction, as $\Sigma \cup H$ contains only $n - 1$ Henkin axioms. Thus we are led to conclude that $\hat{\Sigma}$ is consistent. ■

3.2.1 Exercises

1. Suppose that Σ is inconsistent and ϕ is an $\mathcal{L}$-formula. Prove that $\Sigma \vdash \phi$.

2. Assume that $\Sigma_0 \subseteq \Sigma_1 \subseteq \Sigma_2 \cdots$ are such that each Σ_i is a consistent set of sentences in a language $\mathcal{L}$. Show $\bigcup \Sigma_i$ is consistent.

3. Show that if Π is any consistent set of sentences and σ is a sentence such that $\Pi \cup \{\sigma\}$ is inconsistent, then $\Pi \cup \{\neg\sigma\}$ is consistent. Conclude that in the proof of the Completeness Theorem, if Σ^k is consistent, then Σ^{k+1} is consistent.

4. Prove that the Σ' constructed in the proof of the Completeness Theorem is consistent. [*Suggestion:* Deductions are finite in length.]

5. Prove Lemma 3.2.5.

6. Toward a proof of the Completeness Theorem in a more general setting:

   (a) Do not assume that the language $\mathcal{L}$ is countable. Suppose that you have been given a set of sentences $\Sigma_{\max}$ that is maximal and consistent. So for each sentence σ, either $\sigma \in \Sigma_{\max}$ or $\neg\sigma \in \Sigma_{\max}$. Mimic the proof of Proposition 3.2.6 to convince yourself that we can construct a model $\mathfrak{A}$ of Σ.

   (b) Zorn's Lemma implies the following: If we are given a consistent set of $\mathcal{L}'$-sentences $\hat{\Sigma}$, then the collection of consistent extensions of $\hat{\Sigma}$ has a maximal (with respect to $\subseteq$) element $\Sigma_{\max}$. If you are familiar with Zorn's Lemma, prove this fact.

   (c) Use parts (a) and (b) of this problem to outline a proof of the Completeness Theorem in the case where the language $\mathcal{L}$ is not countable.

7. Complete the proof of the claim on page 94 that the relation $\sim$ is an equivalence relation.

8. Show that the relation $R^{\mathfrak{A}}$ of the structure $\mathfrak{A}$ is well defined. So let R be a relation symbol (a unary relation symbol is fine), and show that if $[t_1] = [t_2]$, then $R^{\mathfrak{A}}([t_1])$ is true if and only if $R^{\mathfrak{A}}([t_2])$ is true.

9. Finish the inductive clause of the proof of Proposition 3.2.6.

10. Fill in the details of the proof of Lemma 3.2.7.

3.3 Compactness

The Completeness Theorem finishes our link between deducibility and logical implication. The Compactness Theorem is our first use of that link. In some sense, what the Compactness Theorem does is focus our attention on the finiteness of deductions, and then we can begin to use that finiteness to our advantage.

Theorem 3.3.1 (Compactness Theorem). *Let Σ be any set of axioms. There is a model of Σ if and only if every finite subset Σ_0 of Σ has a model.*

We say that Σ is **satisfiable** if there is a model of Σ, and we say that Σ is **finitely satisfiable** if every finite subset of Σ has a model. So the Compactness Theorem says that Σ is satisfiable if and only if Σ is finitely satisfiable.

Proof: For the easy direction, suppose that Σ has a model $\mathfrak{A}$. Then $\mathfrak{A}$ is also a model of every finite $\Sigma_0 \subseteq \Sigma$.

For the more difficult direction, assume there is no model of Σ. Then $\Sigma \models \bot$. By the Completeness Theorem, $\Sigma \vdash \bot$, so there is a deduction D of $\bot$ from Σ. Since D is a deduction, it is finite in length and thus can only contain finitely many of the axioms of Σ. Let Σ_0 be the finite set of axioms from Σ that are used in D. Then D is a deduction from Σ_0, so $\Sigma_0 \vdash \bot$. But then by the Soundness Theorem, $\Sigma_0 \models \bot$, so Σ_0 cannot have a model. ■

Corollary 3.3.2. *Let Σ be a set of $\mathcal{L}$-formulas and let θ be an $\mathcal{L}$-formula. $\Sigma \models \theta$ if and only if there is a finite $\Sigma_0 \subseteq \Sigma$ such that $\Sigma_0 \models \theta$.*

Proof:

$\Sigma \models \theta$ iff $\Sigma \vdash \theta$	Soundness and Completeness
iff $\Sigma_0 \vdash \theta$ for a finite $\Sigma_0 \subseteq \Sigma$	deductions are finite
iff $\Sigma_0 \models \theta$	Soundness and Completeness

■

Now we are in a position where we can use the Compactness Theorem to get a better understanding of the limitations of first-order logic—or, to put a more positive spin on it, a better understanding of the richness of mathematics!

Example 3.3.3. Suppose that we examine the structure $\mathfrak{N}$, whose universe is the set of natural numbers $\mathbb{N}$, endowed with the familiar arithmetic functions of addition, multiplication, and exponentiation and the usual binary relation less than. It would be nice to have a collection of axioms that would characterize the structure $\mathfrak{N}$. By this I mean a set of sentences Σ such that $\mathfrak{N} \models \Sigma$, and if $\mathfrak{A}$ is any $\mathcal{L}_{NT}$-structure such that $\mathfrak{A} \models \Sigma$, then $\mathfrak{A}$ is "just like" $\mathfrak{N}$. ($\mathfrak{A}$ is "just like" $\mathfrak{N}$ if there $\mathfrak{A}$ and $\mathfrak{N}$ are isomorphic—see Exercise 5 in Section 1.6.1).

Unfortunately, we cannot hope to have such a set of sentences, and the Compactness Theorem shows us why. Suppose we took any set of sentences Σ that seemed like it ought to characterize $\mathfrak{N}$. Let us add some sentences to Σ and create a new collection of sentences Θ in an extended language $\mathcal{L} = \mathcal{L}_{NT} \cup c$, where c is a new constant symbol:

$$\Theta = \Sigma \cup \{0 < c, S0 < c, SS0 < c, \ldots, \underbrace{SSS\cdots S}_{nS\text{'s}}0 < c, \ldots\}.$$

Now notice that Θ is finitely satisfiable: If Θ_0 is a finite subset of Θ, then Θ_0 is a subset of

$$\Theta_n = \Sigma \cup \{0 < c, S0 < c, SS0 < c, \ldots, \underbrace{SSS\cdots S}_{nS\text{'s}}0 < c\}$$

for some natural number n. But Θ_n has a model $\mathfrak{N}_n$, whose universe is $\mathbb{N}$, the functions and relations are interpreted in the usual way, and $c^{\mathfrak{N}_n} = n+1$. So every finite subset of Θ has a model, and thus Θ has a model $\mathfrak{A}'$. Now forget the interpretation of the constant symbol c and you are left with an $\mathcal{L}_{NT}$-structure $\mathfrak{A} = \mathfrak{A}' \restriction_{\mathcal{L}_{NT}}$. This model $\mathfrak{A}$ is interesting, but we cannot claim that $\mathfrak{A}$ is "just like" $\mathfrak{N}$, since $\mathfrak{A}$ has an element (the thing that used to be called $c^{\mathfrak{A}'}$) such that there are infinitely many elements x that stand in the relation $<$ with that element, while there is no such element of $\mathfrak{N}$. The element $c^{\mathfrak{A}'}$ is called a nonstandard element of the universe, and $\mathfrak{A}$ is another example of a nonstandard model of arithmetic, a model of arithmetic that is not isomorphic to $\mathfrak{N}$. We first encountered nonstandard models of arithmetic in Exercise 7 of Section 2.8.1.

So no set of first-order sentences can completely characterize the natural numbers.

> *Chaff:* Isn't this neat! Notice how each of the $\mathfrak{N}_n$'s in the last example were perfectly ordinary models that looked just like the natural numbers, but the thing that we got at the end looked entirely different!

Definition 3.3.4. If $\mathfrak{A}$ is an $\mathcal{L}$-structure, we define the **theory of** $\mathfrak{A}$ to be $Th(\mathfrak{A}) = \{\phi \mid \phi \text{ is an } \mathcal{L}\text{-formula and } \mathfrak{A} \models \phi\}$. If $\mathfrak{A}$ and $\mathfrak{B}$ are $\mathcal{L}$-structures such that $Th(\mathfrak{A}) = Th(\mathfrak{B})$, then we say that $\mathfrak{A}$ and $\mathfrak{B}$ are **elementarily equivalent**, and write $\mathfrak{A} \equiv \mathfrak{B}$.

If $\mathfrak{A} \equiv \mathfrak{N}$, we say that $\mathfrak{A}$ is a **model of arithmetic**

Example (continued). Notice that the weird structure $\mathfrak{A}$ that we constructed above can be a model of arithmetic if we just let the Σ of our construction be $Th(\mathfrak{N})$. Exercise 2 asks you to prove that in this case we have $\mathfrak{A} \equiv \mathfrak{N}$. Since $\mathfrak{A}$ certainly is not anything like the usual model of arithmetic on the natural numbers, calling $\mathfrak{A}$ a nonstandard model of arithmetic makes pretty good sense. The difficult thing to see is that although the universe A certainly contains nonstandard elements, they don't get in the way of elementary equivalence. The reason for this is that the language $\mathcal{L}_{NT}$ can't refer to any nonstandard element explicitly, so we can't express a statement that is (for example) true in $\mathfrak{N}$ but false in $\mathfrak{A}$. So the lesson to be learned is that it is much easier for two structures to be elementarily equivalent than it is for them to be isomorphic: Our structure $\mathfrak{A}$ is not isomorphic to $\mathfrak{N}$, but $\mathfrak{A}$ is elementarily equivalent to $\mathfrak{N}$.

Example 3.3.5. Remember those ϵ's and δ's from calculus? They were introduced in the nineteenth century in an attempt to firm up the foundations of the subject. When they were developing the calculus, Newton and Leibniz did not worry about limits. They happily used quantities that were infinitely small but not quite zero and they ignored the logical difficulties this presented. These infinitely small quantities live on in today's calculus textbooks as the differentials dx and dy.

Most people find thinking about differentials much easier than fighting through limit computations, and in 1961 Abraham Robinson developed a logical framework for calculus that allowed the use of these infinitesimals in a coherent, noncontradictory way. Robinson's version of the calculus came to be known as nonstandard analysis. Here is a rough introduction (for a complete treatment, see [Keisler 76]).

Taking as our starting point the real numbers that you know so well, we construct a language $\mathcal{L}_{\mathbb{R}}$, the language of the real numbers. For each real number r, the language $\mathcal{L}_{\mathbb{R}}$ includes a constant symbol $\dot{r}$. So the language $\mathcal{L}_{\mathbb{R}}$ includes constant symbols $\dot{0}, \dot{\pi}$, and $\dot{\frac{2}{7}}$. For each function $f : \mathbb{R}^n \to \mathbb{R}$, we toss in a function symbol $\dot{f}$, and for each n-ary relation R on the reals we add an n-ary relation symbol $\dot{R}$. So our language includes, for example, the function symbols $\dot{+}$ and $\dot{\cos}$ and the relation symbol $\dot{<}$.

Now we define $\mathfrak{R}$ to be the $\mathcal{L}_{\mathbb{R}}$-structure $\langle \mathbb{R}, \{r\}, \{f\}, \{R\} \rangle$, where each symbol is interpreted as meaning the number, function, or relation that gave rise to the symbol. So the function symbol $\dot{+}$ stands for the function addition, and the constant symbol $\dot{\pi}$ refers to the real number that is equal to the ratio of the circumference of a circle to its diameter.

Given this structure $\mathfrak{R}$ (notice that $\mathfrak{R}$ is not anything fancy—it is just the real numbers you have been working with since high school), it generates the set of formulas $Th(\mathfrak{R})$, the collection of first-order $\mathcal{L}_{\mathbb{R}}$-formulas that are true statements about the real numbers. Now it is time to use compactness.

Let $\mathcal{L}' = \mathcal{L}_{\mathbb{R}} \cup \{c\}$, where c is a new constant symbol, and look at the collection of $\mathcal{L}'$-sentences

$$\Theta = Th(\mathfrak{R}) \cup \{\dot{0} \dot{<} c\} \cup \{c \dot{<} \dot{r} \mid r \in \mathbb{R}, r > 0\}.$$

(Are you clear about the difference between the dotted and the undotted symbols in this definition?)

By the Compactness Theorem, Θ has a model, $\mathfrak{A}$, and in the model $\mathfrak{A}$, the element denoted by c plays the role of an infinitesimal element: It is positive, yet it is smaller than every positive real number. Speaking roughly, in the universe A of the structure $\mathfrak{A}$ there are three kinds of elements. There are pure standard elements, which constitute a copy of $\mathbb{R}$ that lives inside A. Then there are pure nonstandard elements, for example the element denoted by c. Finally, there are elements such as the object denoted by $\dot{17}\dot{+}c$, which has a standard part and a nonstandard part. (For more of the details, see Exercise 11 in Section 3.4.1.)

The nonstandard elements of the structure $\mathfrak{R}$ provide a method for developing derivatives without using limits. For example, we can define the derivative of a function f at a standard element a to be

$$f'(a) = \text{the standard part of } \frac{f(a+c) - f(a)}{c}.$$

As you can see, there is no limit in the definition. We have traded the limits of calculus for the nonstandard elements of $\mathfrak{A}$, and the slope of a tangent line is nothing more than a slope of a line connecting two points, one of which is not standard. Nonstandard analysis has been an area of active study for the past forty years, and although it is not exactly mainstream, it has been used to discover some new results in various areas of classical analysis.

Example 3.3.6. The idea of coloring a map is supposed to be intuitive. When you were in geography class as a child, you were doubtless given a map of a region and asked to color in the various countries, or states, or provinces. And you were missing the point if you used the same color to shade two countries that shared a common border, although it was permitted to use the same color for two countries whose borders met at a single point. (The states of Utah, Colorado, Arizona, and New Mexico do this in the United States, so coloring both Colorado and Arizona with the color red would be permitted.) The question of how many colors are needed to color any map drawn on the plane was first posed in 1852 by Francis Guthrie, and the answer, that four colors suffice for any such map (as long as each political division consists of a single region—Michigan in a map of the United States or pre-1971 Pakistan in a map of Asia would not be permitted), was proven in 1976 by Kenneth Appel and Wolfgang Hakin. We are not going to prove the Four-Color Theorem here; rather, we extend this result by considering maps with infinitely many regions.

Let R be a set (I'm thinking of the elements of R as being the regions of a map with infinitely many countries) with a symmetric binary relation A (adjacency). Let k be a natural number. We claim that it is possible to assign to each region of R one of k possible colors in such a way that adjacent regions receive different colors if and only if it is possible to so color each finite subset of R.

We will prove this using the Compactness Theorem. One of the tricks to using compactness is to choose your language wisely. For this example, let the language $\mathcal{L}$ consist of a collection of constants $\{r\}_{r \in R}$, one for each region, and a collection of unary predicates $\{C_i\}_{1 \leq i \leq k}$, one for each color. So the atomic statement $C_i(r)$ will be interpreted as meaning that region r gets colored with color i. We will also need a binary relation symbol A, for adjacency.

Let Σ be the collection of sentences:

$$\Sigma = \begin{cases} C_1(r) \vee C_2(r) \vee \cdots \vee C_k(r) & \text{for each } r \in R \\ \neg[C_i(r) \wedge C_j(r)] & r \in R, i \neq j \\ A(r, r') \rightarrow (\neg C_i(r) \wedge C_i(r')) & r, r' \in R, 1 \leq i \leq k \\ A(r, r') & r, r' \in R, r \text{ adjacent to } r' \\ \neg A(r, r') & r, r' \in R, r \text{ not adjacent to } r'. \end{cases}$$

Chaff: Stop now for a minute and make sure that you understand each of the sentences in Σ. You ought to be able to say, in ordinary English, what each sentence asserts. For example, $C_1(r) \vee C_2(r) \vee \cdots \vee C_k(r)$ says that region r must be given one of the k colors. In other words, we have to color each region on the map. Take the time now to translate each of the other statement types of Σ into English.

But now our claim that an infinite map is k-colorable if and only if each finite subset of the map is k-colorable is clear, as a coloring of (a finite subset of) R corresponds to a model of (a finite subset of) Σ, and the Compactness Theorem says that Σ has a model if and only if every finite subset of Σ has a model.

Notice that no quantifiers are used in this example, so we really only needed compactness for predicate logic, not first-order logic. If you are comfortable with the terms, notice also that the proof works whether there are a countably infinite or an uncountably infinite collection of countries.

If you have really been paying attention, you noticed that we did not use the fact that the maps are drawn on the plane. So if we draw a map on a donut with uncountably many countries, it only takes seven colors to color the map, as it was proven in 1890 by Percy John Heawood that seven colors suffice for finite maps drawn on a donut.

Example 3.3.7. You may well be familiar with mathematical trees, as they are often discussed in courses in discrete mathematics or introductory computer science courses. For our purposes a **tree** is a set T partitioned into subsets $T_i, (i = 0, 1, 2, \dots)$, called the levels of the tree, together with a function a such that:

1. T_0 consists of a single element (called the root of the tree).
2. $a : (T - T_0) \to T$ such that if $t \in T_i, i > 0$, then $a(t) \in T_{i-1}$.

A **path** through T consists of a subset $P \subseteq T$ such that $P \cap T_i$ contains exactly one element for each i and P is closed under a. If $t \in T$, the **immediate predecessor** of t is $a(t)$. And an element t_2 is said to be a **predecessor** of t_1 if $t_2 = \underbrace{a(a(\cdots a}_{k\ a\text{'s}}(t_1)))$ for some $k \geq 1$.

We can now use the Compactness Theorem to prove

Lemma 3.3.8 (König's Infinity Lemma). *Let T be a tree all of whose levels are finite and nonempty. Then there is a path through T.*

Proof: Suppose that we are given such a tree T. Let $\mathcal{L}$ be the language consisting of one constant symbol $\hat{t}$ for each element $t \in T$, a unary relation symbol Q, which will be true for elements on the path, and one unary function symbol p, where $p(\hat{t_i})$ is intended to be the immediate predecessor of t_i.

Let Σ be the following set of $\mathcal{L}$-formulas:

$$\Sigma = \begin{cases} p(\hat{t_1}) = \hat{t_2} & \text{for each } t_1, t_2 \in T \text{ such that } a(t_1) = t_2 \\ Q(\hat{t_1}) \vee \cdots \vee Q(\hat{t_k}) & \text{where } T_n = \{t_1, t_2, \dots, t_k\} \text{ (for each } n) \\ \neg(Q(\hat{t_1}) \wedge Q(\hat{t_2})) & \text{for } t_1, t_2 \in T_n, t_1 \neq t_2 \\ Q(\hat{t}) \to Q(p(\hat{t})) & \text{for each } t \in T - T_0. \end{cases}$$

We claim that Σ is finitely satisfiable: Let Σ_0 be a finite subset of Σ, and let n be so large that if $\hat{t}$ is mentioned in Σ_0, then $t \in T_0 \cup T_1 \cup \cdots \cup T_n$. Pick any element $t^* \in T_{n+1}$, and build an $\mathcal{L}$-structure $\mathfrak{A}$ by letting the universe A be the tree T, $\hat{t}^{\mathfrak{A}}$ be t, letting $p^{\mathfrak{A}}$ be the function a, and letting $Q^{\mathfrak{A}}$ be the collection of predecessors of t^*. It is easy to check that $\mathfrak{A}$ is a model of Σ_0, and thus by compactness, there is a structure $\mathfrak{B}$ such that $\mathfrak{B}$ is a model of Σ. If we let $P = \{t \in T \mid \hat{t}^{\mathfrak{B}} \in Q^{\mathfrak{B}}\}$, then P is a path through T, and König's Infinity Lemma is proven. ■

3.3.1 Exercises

1. A common attempt to try to write a set of axioms that would characterize $\mathfrak{N}$ (see Example 3.3.3) is to let Σ be the collection of *all* $\mathcal{L}_{NT}$-formulas that are true in $\mathfrak{N}$, and then to argue that this is an element of Σ:

$$(\forall x)(\exists n)(x = \underbrace{SSS \cdots S}_{n S\text{'s}} 0).$$

Therefore, there can be no nonstandard elements in any model of Σ. Explain why this reasoning fails.

2. Show that if we let $\Sigma = Th(\mathfrak{N})$ in the construction of Example 3.3.3, then the structure $\mathfrak{A}$ that is constructed is elementarily equivalent to the structure $\mathfrak{N}$. Thus $\mathfrak{A}$ is a model of arithmetic.

3. Show that if $\mathfrak{A}$ and $\mathfrak{B}$ are $\mathcal{L}$-structures such that $\mathfrak{A} \cong \mathfrak{B}$, then $\mathfrak{A} \equiv \mathfrak{B}$.

4. Suppose that Σ is a set of $\mathcal{L}$-sentences such that at least one sentence from Σ is true in each $\mathcal{L}$-structure. Show that the disjunction of some finitely many sentences from Σ is logically valid.

5. Show that every nonstandard model of arithmetic contains an infinite prime number, that is, an infinite number a such that if $a = bc$, then either $b = 1$ or $c = 1$.

6. Show that if $\phi(x)$ is a formula with one free variable in $\mathcal{L}_{NT}$ such that there are infinitely many natural numbers a such that $\mathfrak{N} \models \phi(x)[s[x|a]]$, then in every nonstandard model of arithmetic there is an infinite number b such that $\mathfrak{N} \models \phi(x)[s[x|b]]$.

7. Verify that we can use the Compactness Theorem in Example 3.3.5 by verifying that every finite subset of Θ has a model.

8. (a) Using only connectives, quantifiers, variables, and the equality symbol, construct a set of sentences Σ such that every model of Σ is infinite.

(b) Prove that if Γ is a set of sentences with arbitrarily large finite models, then Γ has an infinite model.

(c) Show that there can be no set of sentences in first-order logic that characterizes the finite groups. (See Exercise 3 in Section 2.8.1.)

(d) Prove that there is no finite set of sentences

$$\Phi = \{\phi_1, \phi_2, \dots, \phi_n\}$$

such that $\mathfrak{A} \models \Phi$ if and only if A is infinite. [*Suggestion:* Look at $\neg(\phi_1 \wedge \phi_2 \wedge \cdots \wedge \phi_n)$.]

9. Suppose that Σ_1 and Σ_2 are two sets of sentences such that no structure is a model of both Σ_1 and Σ_2. Show there is a sentence α such that every model of Σ_1 is also a model of α and furthermore, every model of Σ_2 is a model of $\neg\alpha$.

10. A binary relation $<$ on a set A is said to be a **linear order** if

(a) $<$ is irreflexive—$(\forall a \in A)(\neg a < a)$.

(b) $<$ is transitive—$(\forall a, b, c \in A)\big([a < b \wedge b < c] \rightarrow a < c\big)$.

(c) $<$ satisfies trichotomy—$\forall a, b \in A$ exactly one of the following is true: $a < b, b < a$, or $a = b$.

If a linear order $<$ has the additional property that there are no infinite descending chains—there do not exist $a_1, a_2, \ldots \in A$ such that $a_1 > a_2 > a_3 > \cdots$ (where $a_1 > a_2$ means $a_2 < a_1$), then the relation $<$ is a **well-order** of the set A. Suppose that $\mathcal{L}$ is a language containing a binary relation symbol $<$. Show there is no set of $\mathcal{L}$-sentences Σ such that Σ has both of the following properties:

(a) Σ has an infinite model $\mathfrak{A}$ in which $<^{\mathfrak{A}}$ is a linear order of A.

(b) If $\mathfrak{B}$ is any infinite model of Σ, then $<^{\mathfrak{B}}$ is a well-ordering of B.

11. Show that $<$ is not a well-order in any nonstandard model of arithmetic.

12. (a) In the structure $\mathfrak{A}$ that was built in Example 3.3.5, explain how we know that

$$\mathfrak{A} \models (\forall x)\big[(x \dot{>} \dot{0}) \rightarrow (x \dot{/} \dot{2} \dot{>} \dot{0} \wedge x \dot{>} x \dot{/} \dot{2})\big].$$

(b) Show that $<$ is a linear order of A, the universe of $\mathfrak{A}$.

(c) Show that $<$ is not a well-order in this structure.

3.4 Substructures and the Löwenheim–Skolem Theorems

In this section we will discuss a relation between structures. A given set of sentences may have many different models, and it will turn out that in some cases those models are related in surprising ways. We begin by defining the notion of a substructure.

Definition 3.4.1. If $\mathfrak{A}$ and $\mathfrak{B}$ are two $\mathcal{L}$-structures, we will say that $\mathfrak{A}$ is a **substructure** of $\mathfrak{B}$, and write $\mathfrak{A} \subseteq \mathfrak{B}$, if:

1. $A \subseteq B$.

2. For every constant symbol c, $c^{\mathfrak{A}} = c^{\mathfrak{B}}$.

3. For every n-ary relation symbol R, $R^{\mathfrak{A}} = R^{\mathfrak{B}} \cap A^n$.

4. For every n-ary function symbol f, $f^{\mathfrak{A}} = f^{\mathfrak{B}} \restriction_{A^n}$. In other words, for every n-ary function symbol f and every $a \in A$, $f^{\mathfrak{A}}(a) = f^{\mathfrak{B}}(a)$. (This is called the **restriction of the function $f^{\mathfrak{A}}$ to the set A^n**.)

Thus a substructure of $\mathfrak{B}$ is completely determined by its universe, and this universe can be any nonempty subset of B that contains the constants and is closed under every function f.

Example 3.4.2. Suppose that we try to build a substructure $\mathfrak{A}$ of the structure $\mathfrak{N} = (\mathbb{N}, 0, S, +, \cdot, E, <)$. Since A must be closed under the functions and contain the constants, the number 0 must be an element of the universe A. But now, since the substructure must be closed under the function S, it is clear that every natural number must be an element of A. Thus $\mathfrak{N}$ has no proper substructures.

Example 3.4.3. Now, suppose that we try to find some substructures of the structure $\mathfrak{B} = \langle \mathbb{N}, 0, < \rangle$, with the usual interpretations of 0 and $<$. Since there are no function symbols, any nonempty subset of $\mathbb{N}$ that includes the number 0 can serve as the universe of a substructure $\mathfrak{A} \subseteq \mathfrak{B}$.

Suppose that we let $\mathfrak{A} = \langle \{0\}, 0, < \rangle$. Then notice that even though $\mathfrak{A} \subseteq \mathfrak{B}$, there are plenty of sentences that are true in one structure that are not true in the other structure. For example, $(\forall x)(\exists y)x < y$ is false in $\mathfrak{A}$ and true in $\mathfrak{B}$. It will not be hard for you to find an example of a sentence that is true in $\mathfrak{A}$ and false in $\mathfrak{B}$.

As Example 3.4.3 shows, if we are given two structures such that $\mathfrak{A} \subseteq \mathfrak{B}$, most of the time you would expect that $\mathfrak{A}$ and $\mathfrak{B}$ would be very different, and there would be lots of sentences that would be true in one of the structures that would not be true in the other.

Sometimes, however, truth in the smaller structure is more closely tied to truth in the larger structure.

Definition 3.4.4. Suppose that $\mathfrak{A}$ and $\mathfrak{B}$ are $\mathcal{L}$-structures and $\mathfrak{A} \subseteq \mathfrak{B}$. We say that $\mathfrak{A}$ is an **elementary substructure** of $\mathfrak{B}$ (equivalently, $\mathfrak{B}$ is an **elementary extension** of $\mathfrak{A}$), and write $\mathfrak{A} \prec \mathfrak{B}$, if for every $s : Vars \to A$ and for every $\mathcal{L}$-formula ϕ,

$$\mathfrak{A} \models \phi[s] \text{ if and only if } \mathfrak{B} \models \phi[s].$$

Chaff: Notice that if we want to prove $\mathfrak{A} \prec \mathfrak{B}$, we need only prove $\mathfrak{A} \models \phi[s] \to \mathfrak{B} \models \phi[s]$, since once we have done that, the other direction comes for free by using the contrapositive and negations.

Proposition 3.4.5. *Suppose that $\mathfrak{A} \prec \mathfrak{B}$. Then a sentence σ is true in $\mathfrak{A}$ if and only if it is true in $\mathfrak{B}$.*

Proof: Exercise 5. ■

Example 3.4.6. We saw earlier that the structure $\mathfrak{B} = \langle \mathbb{N}, 0, < \rangle$ has lots of substructures. However, $\mathfrak{B}$ has no proper elementary substructures. For suppose that $\mathfrak{A} \prec \mathfrak{B}$. Certainly, $0 \in A$, as $\mathfrak{A}$ is a substructure. Since the sentence $(\exists y)\big[0 < y \wedge (\forall x)(0 < x \rightarrow y \leq x)\big]$ is true in $\mathfrak{B}$, it must be true in $\mathfrak{A}$ as well. So

$$\mathfrak{A} \models (\exists y)\big[0 < y \wedge (\forall x)(0 < x \rightarrow y \leq x)\big].$$

Thus, for any assignment function $s : Vars \rightarrow A$ there is some $a \in A$ such that

$$\mathfrak{A} \models \big[0 < y \wedge (\forall x)(0 < x \rightarrow y \leq x)\big][s[y|a]].$$

Fix such an s and such an $a \in A$. Now we use elementarity again. Since $\mathfrak{A} \prec \mathfrak{B}$ and $s[y|a] : Vars \rightarrow A$, we know that

$$\mathfrak{B} \models \big[0 < y \wedge (\forall x)(0 < x \rightarrow y \leq x)\big][s[y|a]].$$

But in the structure $\mathfrak{B}$, there is a unique element that makes the formula $\big[0 < y \wedge (\forall x)(0 < x \rightarrow y \leq x)\big]$ true, namely the number 1. So a must be the number 1, and so 1 must be an element of A. Similarly, you can show that $2 \in A$, $3 \in A$, and so on. Thus $\mathbb{N} \subseteq A$, and $\mathfrak{A}$ will not be a proper elementary substructure of $\mathfrak{B}$.

This example shows that when building an elementary substructure of a given structure $\mathfrak{B}$, we need to make sure that witnesses for each existential sentence true in $\mathfrak{B}$ must be included in the universe of the elementary substructure $\mathfrak{A}$. That idea will be the core of the proof of the Downward Löwenheim–Skolem Theorem, Theorem 3.4.8. In fact, the next lemma says that making sure that such witnesses are elements of $\mathfrak{A}$ is all that is needed to ensure that $\mathfrak{A}$ is an elementary substructure of $\mathfrak{B}$.

Lemma 3.4.7. *Suppose that $\mathfrak{A} \subseteq \mathfrak{B}$ and that for every formula α and every $s : Vars \rightarrow A$ such that $\mathfrak{B} \models \exists x \alpha[s]$ there is an $a \in A$ such that $\mathfrak{B} \models \alpha[s[x|a]]$. Then $\mathfrak{A} \prec \mathfrak{B}$.*

Proof: We will show, given the assumptions of the lemma, that if ϕ is any formula and s is any variable assignment function into A, $\mathfrak{A} \models \phi[s]$ if and only if $\mathfrak{B} \models \phi[s]$, and thus $\mathfrak{A} \prec \mathfrak{B}$.

This is an easy proof by induction on the complexity of ϕ, which we will make even easier by noting that we can replace the $\forall$ inductive step by an $\exists$ inductive step, as $\forall$ can be defined in terms of $\exists$.

So for the base case, assume that ϕ is atomic. For example, if ϕ is $R(x, y)$, then $\mathfrak{A} \models \phi[s]$ if and only if $\langle s(x), s(y)\rangle \in R^{\mathfrak{A}}$. But

$R^{\mathfrak{A}} = R^{\mathfrak{B}} \cap A^2$, so $\langle s(x), s(y)\rangle \in R^{\mathfrak{A}}$ if and only if $\langle s(x), s(y)\rangle \in R^{\mathfrak{B}}$. But $\langle s(x), s(y)\rangle \in R^{\mathfrak{B}}$ if and only if $\mathfrak{B} \models \phi[s]$, as needed.

For the inductive clauses, assume that ϕ is $\neg\alpha$. Then

$$
\begin{aligned}
\mathfrak{A} \models \phi[s] & \text{ if and only if } \mathfrak{A} \models \neg\alpha[s] \\
& \text{ if and only if } \mathfrak{A} \not\models \alpha[s] \\
& \text{ if and only if } \mathfrak{B} \not\models \alpha[s] \quad \text{inductive hypothesis} \\
& \text{ if and only if } \mathfrak{B} \models \neg\alpha[s] \\
& \text{ if and only if } \mathfrak{B} \models \phi[s].
\end{aligned}
$$

The second inductive clause, if ϕ is $\alpha \vee \beta$, is similar.

For the last inductive clause, suppose that ϕ is $\exists x\alpha$. Suppose also that $\mathfrak{A} \models \phi[s]$; in other words, $\mathfrak{A} \models \exists x\alpha[s]$. Then, for some $a \in A$, $\mathfrak{A} \models \alpha[s[x|a]]$. Since $s[x|a]$ is a function mapping variables into A, by our inductive hypothesis, $\mathfrak{B} \models \alpha[s[x|a]]$. But then $\mathfrak{B} \models \exists x\alpha[s]$, as needed. For the other direction, assume that $\mathfrak{B} \models \exists x\alpha[s]$, where $s : \mathit{Vars} \to A$. We use the assumption of the lemma to find an $a \in A$ such that $\mathfrak{B} \models \alpha[s[x|a]]$. As $s[x|a]$ is a function with codomain A, by the inductive hypothesis $\mathfrak{A} \models \alpha[s[x|a]]$, and thus $\mathfrak{A} \models \exists x\alpha[s]$, and the proof is complete. ■

Chaff: We are now going to look at the Löwenheim–Skolem Theorems, which were published in 1915. To understand these theorems, you need to have at least a basic understanding of cardinality, a topic that is outlined in the Appendix. However, if you are in a hurry, it will suffice if you merely remember that there are many different sizes of infinite sets. An infinite set A is countable if there is a bijection between A and the set of natural numbers $\mathbb{N}$ otherwise, the set is uncountable. Examples of countable sets include the integers and the set of rational numbers. The set of real numbers is uncountable, in that there is no bijection between $\mathbb{R}$ and $\mathbb{N}$. So there are more reals than natural numbers. There are infinitely many different sizes of infinite sets. The smallest infinite size is countable.

Theorem 3.4.8 (Downward Löwenheim–Skolem Theorem). *Suppose that $\mathcal{L}$ is a countable language and $\mathfrak{B}$ is an $\mathcal{L}$-structure. Then $\mathfrak{B}$ has a countable elementary substructure.*

Proof: If B is finite or countably infinite, then $\mathfrak{B}$ is its own countable elementary substructure, so assume that B is uncountable. As the language $\mathcal{L}$ is countable, there are only countably many $\mathcal{L}$-formulas, and thus only countably many formulas of the form $\exists x\alpha$.

Let A_0 be any nonempty countable subset of B. We show how to build A_1 such that $A_0 \subseteq A_1$, and A_1 is countable. The idea is to add to A_0 witnesses for the truth (in $\mathfrak{B}$) of existential statements.

Notice that as A_0 is countable, there are only countably many functions $s' : Vars \to A_0$ that are eventually constant. (This is a nice exercise for those of you who have had a course in set theory or are reasonably comfortable with cardinality arguments.) Also, if we are given any ϕ and any $s : Vars \to A_0$, we can find an eventually constant $s' : Vars \to A_0$ such that s and s' agree on the free variables of ϕ, and thus $\mathfrak{B} \models \phi[s]$ if and only if $\mathfrak{B} \models \phi[s']$.

The construction of A_1*:* For each formula of the form $\exists x\alpha$ and each $s : Vars \to A_0$ such that $\mathfrak{B} \models \exists x\alpha[s]$, find an eventually constant $s' : Vars \to A_0$ such that s and s' agree on the free variables of $\exists x\alpha$. Pick an element $a_{\alpha,s'} \in B$ such that $\mathfrak{B} \models \alpha[s[x|a_{\alpha,s'}]]$, and let

$$A_1 = A_0 \cup \{a_{\alpha,s'}\}_{\text{all } \alpha, s:\, Vars \to A_0}.$$

Notice that A_1 is countable, as there are only countably many α's and countably many s'.

Continue this construction, iteratively building A_{n+1} from A_n. Let $A = \cup_{n=0}^{\infty} A_n$. As A is a countable union of countable sets, A is countable.

Now we have constructed a potential universe A for a substructure for $\mathfrak{B}$. We have to prove that A is closed under the functions of $\mathfrak{B}$ (by the remarks following Definition 3.4.1 this shows that $\mathfrak{A}$ is a substructure of $\mathfrak{B}$), and we have to show that $\mathfrak{A}$ satisfies the criteria set out in Lemma 3.4.7, so we will know that $\mathfrak{A}$ is an elementary substructure of $\mathfrak{B}$.

First, to show that A is closed under the functions of $\mathfrak{B}$, suppose that $a \in A$ and f is a unary function symbol (the general case is identical) and that $b = f^{\mathfrak{B}}(a)$. We must show that $b \in A$. Fix an n so large that $a \in A_n$, let ϕ be the formula $(\exists y) y = f(x)$, and let s be any assignment function into A such that $s(x) = a$. We know that $\mathfrak{B} \models (\exists y) y = f(x)[s]$, and we know that if $\mathfrak{B} \models \big(y = f(x)\big)[s[y|d]]$, then $d = b$. So, in our construction of A_{n+1} we must have used $a_{y=f(x),s} = b$, so $b \in A_{n+1}$, and $b \in A$, as needed.

In order to use Lemma 3.4.7, we must show that if α is a formula and $s : Vars \to A$ is such that $\mathfrak{B} \models \exists x\alpha[s]$, then there is an $a \in A$ such that $\mathfrak{B} \models \alpha[s[x|a]]$. So, fix such an α and such an s. Find an eventually constant $s' : Vars \to A$ such that s and s' agree on all the free variables of α. Thus $\mathfrak{B} \models \exists x\alpha[s']$, and all of the values of s' are elements of some fixed A_n, as s' takes on only a finite number of values. But then by construction of A_{n+1}, there is an element a of A_{n+1} such that $\mathfrak{B} \models \alpha[s'[x|a]]$. But this tells us (since s and s' agree on the free variables of α) that $\mathfrak{B} \models \alpha[s[x|a]]$, as needed.

So we have met the hypotheses of Lemma 3.4.7, and thus $\mathfrak{A}$ is a countable elementary substructure of $\mathfrak{B}$, as needed. ■

Chaff: I would like to look at a bit of this proof a little more closely. In the construction of A_1, what we did was to find an $a_{\alpha,s'}$ for each formula $\exists x\alpha$ and each $s : Vars \to A$, and the point was that $a_{\alpha,s'}$ would be a witness to the truth in $\mathfrak{B}$ of the existential statement $\exists x\alpha$. So we have constructed a *function* which, given an existential formula $\exists x\alpha$ and an assignment function, finds a value for x that makes the formula α true. A function of this sort is called a **Skolem function**, and the construction of A in the proof of the Downward Löwenheim–Skolem Theorem can thus be summarized: Let A_0 be a countable subset of B, and form the closure of A_0 under the set of all Skolem functions. Then show that this closure is an elementary substructure of $\mathfrak{B}$.

Example 3.4.9. We saw an indication in Exercise 4 in Section 2.8.1 that the axioms of Zermelo–Fraenkel set theory (known as ZF) can be formalized in first-order logic. Accepting that as true (which it is), we know that if the axioms are consistent they have a model, and then by the Downward Löwenheim–Skolem Theorem, there must be a countable model for set theory. But this is interesting, as the following are all theorems of ZF:

- There is a countably infinite set.
- If a set a exists, then the collection of subsets of a exists.
- If a is countably infinite, then the collection of subsets of a is uncountable. (This is Cantor's Theorem).

Now, let us suppose that $\mathfrak{A}$ is our countable model of ZF, and suppose that a is an element of A and is countably infinite. If b is the set of all of the subsets of a, we know that b is uncountable (by Cantor's Theorem) and yet b must be countable, as all of the elements of b are in the model $\mathfrak{A}$, and $\mathfrak{A}$ is countable! So b must be both countable and uncountable! This is called (somewhat incorrectly) Skolem's paradox, and Exercise 8 asks you to figure out the solution to the paradox.

Probably the way to think about the Downward Löwenheim–Skolem Theorem is that it guarantees that if there are any infinite models of a given set of formulas, then there is a small (countably infinite means small) model of that set of formulas. It seems reasonable to ask if there is a similar guarantee about big models, and there is.

Proposition 3.4.10. *Suppose that Σ is a set of $\mathcal{L}$-formulas with an infinite model. If κ is an infinite cardinal, then there is a model of Σ of cardinality greater than or equal to κ.*

Proof: This is an easy application of the Compactness Theorem. Expand $\mathcal{L}$ to include κ new constant symbols c_i, and let $\Gamma = \Sigma \cup \{c_i \neq c_j \mid i \neq j\}$. Then Γ is finitely satisfiable, as we can take our given infinite model of Σ and interpret the c_i in that model in such a way that $c_i \neq c_j$ for any finite set of constant symbols. By the Compactness Theorem, there is a structure $\mathfrak{A}$ that is a model of Γ, and thus certainly the cardinality of A is greater than or equal to κ. If we restrict $\mathfrak{A}$ to the original language, we get a model of Σ of the required cardinality. ■

Corollary 3.4.11. *If Σ is a set of formulas from a countable language with an infinite model, and if κ is an infinite cardinal, then there is a model of Σ of cardinality κ.*

Proof: First, use Proposition 3.4.10 to get $\mathfrak{B}$, a model of Σ of cardinality greater than or equal to κ. Then, mimic the proof of the Downward Löwenheim–Skolem Theorem, starting with a set $A_0 \subseteq B$ of cardinality exactly κ. Then the A that is constructed in that proof also will have cardinality κ, and as $\mathfrak{A} \prec \mathfrak{B}$, $\mathfrak{A}$ will be a model of Σ of cardinality κ. ■

Corollary 3.4.12. *If $\mathfrak{A}$ is an infinite $\mathcal{L}$-structure, then there is* no *set of first-order formulas that characterize $\mathfrak{A}$ up to isomorphism.*

Proof: More precisely, the corollary says that there is no set of formulas Σ such that $\mathfrak{B} \models \Sigma$ if and only if $\mathfrak{A} \cong \mathfrak{B}$. We know that there are models of Σ of all cardinalities, and we know that there are no bijections between sets of different cardinalities. So there must be many models of Σ that are not isomorphic to $\mathfrak{A}$. ■

> *Chaff:* There are sets of axioms that do characterize infinite structures. For example, the second-order axioms of Peano Arithmetic include axioms to ensure that addition and multiplication behave normally, and they also include the principle of mathematical induction: If M is a set of numbers, if $0 \in M$, and if $S(n) \in M$ for every n such that $n \in M$, then $(\forall n)(n \in M)$.
>
> Any model of Peano Arithmetic is isomorphic to the natural numbers, but notice that we used two notions (sets of numbers and the elementhood relation) that are not part of our description of $\mathfrak{N}$. By introducing sets of numbers we have left the world of first-order logic and have entered second-order logic, and it is only by using second-order logic that we are able to characterize $\mathfrak{N}$. For a nice discussion of this topic, see [Bell and Machover 77, Chapter 7, Section 2].

The results from Proposition 3.4.10 to Corollary 3.4.12 give us models that are large, but they have a slightly different flavor from the Downward Löwenheim–Skolem Theorem, in that they do not guarantee that the small model is an elementary substructure of the large model. That is the content of the Upward Löwenheim–Skolem Theorem, a proof of which is outlined in the Exercises.

Theorem 3.4.13 (Upward Löwenheim–Skolem Theorem). *If $\mathcal{L}$ is a countable language, $\mathfrak{A}$ is an infinite $\mathcal{L}$-structure, and κ is a cardinal, then $\mathfrak{A}$ has an elementary extension $\mathfrak{B}$ such that the cardinality of B is greater than or equal to κ.*

3.4.1 Exercises

1. Suppose that $\mathfrak{B} \subseteq \mathfrak{A}$, that ϕ is of the form $(\forall x)\psi$, where ψ is quantifier-free, and that $\mathfrak{A} \models \phi$. Prove that $\mathfrak{B} \models \phi$. The short version of this fact is, "Universal sentences are preserved downward." Formulate and prove the corresponding fact for existential sentences.

2. Justify the *Chaff* following Definition 3.4.4.

3. Show that if $\mathfrak{A} \prec \mathfrak{B}$ and $\mathfrak{C} \prec \mathfrak{B}$ and $\mathfrak{A} \subseteq \mathfrak{C}$, then $\mathfrak{A} \prec \mathfrak{C}$.

4. Suppose that we have an **elementary chain**, a set of $\mathcal{L}$-structures such that

$$\mathfrak{A}_1 \prec \mathfrak{A}_2 \prec \mathfrak{A}_3 \prec \cdots$$

and let $\mathfrak{A} = \bigcup_{i=1}^{\infty} \mathfrak{A}_i$. So the universe A of $\mathfrak{A}$ is the union of the universes A_i, $R^{\mathfrak{A}} = \bigcup_{i=1}^{\infty} R^{\mathfrak{A}_i}$, etc. Show that $\mathfrak{A}_i \prec \mathfrak{A}$ for each i. [*Suggestion:* To show that $\mathfrak{A}_i \subseteq \mathfrak{A}$ is pretty easy by the definition. To get that $\mathfrak{A}$ is an *elementary* extension, you have to use induction on formulas. Notice by the comments following Definition 3.4.4 that you need only prove one direction. You may find it easier to use $\exists$ rather than $\forall$ in the quantifier part of the inductive step of the proof.]

5. Prove Proposition 3.4.5.

6. Show that if $\mathfrak{A} \prec \mathfrak{B}$ and if there is an element $b \in B$ and a formula $\phi(x)$ such that $\mathfrak{B} \models \phi[s[x|b]]$ and for every other $\hat{b} \in B$, $\mathfrak{B} \not\models \phi[s[x|\hat{b}]]$, then $b \in A$. [*Suggestion:* This is very similar to Example 3.4.6].

7. Suppose that $\mathfrak{B} = \{\mathbb{N}, +, \cdot\}$, and let $A_0 = \{2, 3\}$. Let F be the set of Skolem functions $\{f_{\alpha,s}\}$ corresponding to αs of the form $(\exists x)x = yz$. Find the closure of A_0 under F. [*Suggestion:* Do not forget that the assignment functions s that you need to consider are functions mapping into A_0 at first, then A_1, and so on. You probably want to explicitly write out A_1, then A_2, etc. I am using the notation here corresponding to the proof of Theorem 3.4.8.]

8. To say that a set a is countable means that there is a function with domain the natural numbers and codomain a that is a bijection. Notice that this is an existential statement, saying that a certain kind of function *exists*. Now, think about Example 3.4.9 and see if you can figure out why it is not really a contradiction that the set b is both countable and uncountable. In particular, think about what it means for an existential statement to be true in a structure $\mathfrak{A}$, as opposed to true in the real world (whatever *that* means!)

9. (Toward the Proof of the Upward Löwenheim–Skolem Theorem) If $\mathfrak{A}$ is an $\mathcal{L}$-structure, let $\mathcal{L}(A) = \mathcal{L} \cup \{\overline{a} \mid a \in A\}$, where each $\overline{a}$ is a new constant symbol. Then, let $\overline{\mathfrak{A}}$ be the $\mathcal{L}(A)$-structure having the same universe as $\mathfrak{A}$ and the same interpretation of the symbols of $\mathcal{L}$ as $\mathfrak{A}$, and interpreting each $\overline{a}$ as a. Then we define the **complete diagram of** $\mathfrak{A}$ as

$$Th(\overline{\mathfrak{A}}) = \{\sigma \mid \sigma \text{ is an } \mathcal{L}(A)\text{-formula such that } \overline{\mathfrak{A}} \models \sigma\}.$$

Show that if $\overline{\mathfrak{B}}$ is any model of $Th(\overline{\mathfrak{A}})$, and if $\mathfrak{B} = \overline{\mathfrak{B}}\restriction_{\mathcal{L}}$, then $\mathfrak{A}$ is isomorphic to an elementary substructure of $\mathfrak{B}$. [*Suggestion:* Let $h : A \to B$ be given by $h(a) = \overline{a}^{\overline{\mathfrak{B}}}$. Let C be the range of h. Show C is closed under $f^{\mathfrak{B}}$ for every f in $\mathcal{L}$, and thus C is the universe of $\mathfrak{C}$, a substructure of $\mathfrak{B}$. Then show h is an isomorphism between $\mathfrak{A}$ and $\mathfrak{C}$. Finally, show that $\mathfrak{C} \prec \mathfrak{B}$.]

10. Use Exercise 9 to prove the Upward Löwenheim–Skolem Theorem by finding a model $\overline{\mathfrak{B}}$ of the complete diagram of the given model $\mathfrak{A}$ such that the cardinality of $\overline{B}$ is greater than or equal to κ.

11. We can now fill in some of the details of our discussion of nonstandard analysis from Example 3.3.5. As the language $\mathcal{L}_{\mathbb{R}}$ of that example already includes constant symbols for each real number, the complete diagram of $\mathfrak{R}$ is nothing more than $Th(\mathfrak{R})$. Explain how Exercise 9 shows that there is an isomorphic copy of the real line living inside the structure $\mathfrak{A}$.

3.5 Summing Up, Looking Ahead

We have proven a couple of difficult theorems in this chapter, and by understanding the proof of the Completeness Theorem you have grasped an intricate argument with a wonderful idea at its core. Our results have been directed at structures: What kinds of structures exist? How can we (or can't we) characterize them? How large can they be?

The next chapter begins our discussion of Kurt Gödel's famous incompleteness theorems. Rather than discussing the strength of our deductive system as we have done in the last two chapters, we will now discuss the strength of sets of axioms. In particular, we will look at the question of how complicated a set of axioms must be in order to prove all of the true statements about the standard structure $\mathfrak{N}$.

In Chapter 4 we will introduce the idea of coding up the statements of $\mathcal{L}_{NT}$ as terms and will show that a certain set of nonlogical axioms is strong enough to prove some basic facts about the numbers coding up those statements. Then, in Chapter 5, we will bring those facts together to show that the expressive power we have gained has allowed us to express truths that are unprovable from our set of axioms.

Chapter 4

Incompleteness—Groundwork

4.1 Introduction

Now, I hope you have been paying attention closely enough to be bothered by the title of this chapter. The preceding chapter was about completeness, and we proved the Completeness Theorem. Now we seem to be launching an investigation of incompleteness! This point is pretty confusing, so let us try to start out as clearly as possible.

In Chapter 3 we proved the completeness of our axiomatic system. We have shown that the deductive system described in Chapter 2 is sound and complete. What does this mean? For the collection of logical axioms and rules of inference that we have set out, any formula ϕ that can be deduced from Σ will be true in all models of Σ under any variable assignment function (that's soundness), and furthermore any formula ϕ that is true in all models of Σ under every assignment function will be deducible from Σ (that's completeness). Thus, our deductive system is as nice as it can possibly be. The rough version of the Completeness and Soundness Theorems is: We can prove it if and only if it is true everywhere.

Now we will change our focus. Rather than discussing the wonderful qualities of our deductive system, we will concentrate on a particular language, $\mathcal{L}_{NT}$, and think about a particular structure: $\mathfrak{N}$, the natural numbers.

Wouldn't life be just great if we knew that we could prove every true statement about the natural numbers? Of course, the statements that we can prove depend on our choice of nonlogical axioms Σ, so let me start this paragraph over.

Wouldn't life be just great if we could find a set of nonlogical axioms that could prove every true statement about the natural numbers? I would love to have a set of axioms Σ such that $\mathfrak{N} \models \Sigma$ (so our axioms are true statements about the natural numbers) and Σ is rich enough so that for every sentence σ, if $\mathfrak{N} \models \sigma$, then $\Sigma \vdash \sigma$. Since Σ has a model, we know that Σ is consistent, so by soundness my wished-for Σ will prove exactly those sentences that are true in $\mathfrak{N}$. The set of sentences of $\mathcal{L}_{NT}$ that are true in $\mathfrak{N}$ is called the **Theory of** $\mathfrak{N}$, or $Th(\mathfrak{N})$.

Since we know that a sentence is either true in $\mathfrak{N}$ or false in $\mathfrak{N}$, this set of axioms Σ is complete—complete in the sense that given any sentence σ, Σ will provide either a deduction of σ or a deduction of $\neg\sigma$.

Definition 4.1.1. A set of nonlogical axioms Σ is called **complete** if for every sentence σ, either $\Sigma \vdash \sigma$ or $\Sigma \vdash \neg\sigma$.

> *Chaff:* To reiterate, in Chapter 3 we showed that our *deductive system* is complete. This means that for a given Σ, the deductive system will prove exactly those formulas that are logical consequences of Σ. When we say that a set of *axioms* is complete, we are saying that the axioms are strong enough to provide either a proof or a refutation of any sentence. This is harder.

Our goal is to find a complete and consistent set of axioms Σ such that $\mathfrak{N} \models \Sigma$. So this set Σ would be strong enough to prove every $\mathcal{L}_{NT}$-sentence that is true in the standard structure $\mathfrak{N}$. Such a set of axioms is said to axiomatize $Th(\mathfrak{N})$.

Definition 4.1.2. A set of axioms Σ is an **axiomatization of** $Th(\mathfrak{N})$ if for every sentence $\sigma \in Th(\mathfrak{N})$, $\Sigma \vdash \sigma$.

Actually, as stated, it is pretty easy to find an axiomatization of $Th(\mathfrak{N})$: Just let the axiom set be $Th(\mathfrak{N})$ itself. This set clearly axiomatizes itself, so we are finished! Off we go to have a drink. Of course, our answer to the search has the problem that we don't have an easy way to tell exactly which formulas are elements of the set of axioms. If I took a random sentence in $\mathcal{L}_{NT}$ and asked you if this sentence were true in the standard structure, I doubt you'd be able to tell me. The truth of nonrandom sentences is also hard to figure out—consider the twin prime conjecture, that there are infinitely many pairs of numbers k and $k+2$ such that both k and $k+2$ are prime. People have been thinking about that one for over 2000 years and we don't know if it is true or not. So we have no idea if the twin prime conjecture is in $Th(\mathfrak{N})$ or not. So it looks like $Th(\mathfrak{N})$ is unsatisfactory as a set of nonlogical axioms.

So, to refine our question a bit, what we would like is a set of nonlogical axioms Σ that is simple enough so that we can recognize whether or not a given formula is an axiom (so the set of axioms should be decidable) and strong enough to prove every formula in $Th(\mathfrak{N})$. So we search for a complete, consistent, decidable set of axioms for $\mathfrak{N}$. Unfortunately, our search is doomed to failure, and that fact is the content of Gödel's First Incompleteness Theorem, which we shall prove in Chapter 5.

There is a fair bit of groundwork to cover before we get to the Incompleteness Theorem, and much of that groundwork is rather technical. Here is a thumbnail sketch of our plan to reach the theorem: The proof of the First Incompleteness Theorem essentially consists of constructing a certain sentence θ and noticing that θ is, by its very nature, a true statement in $\mathfrak{N}$ and a statement that is unprovable from our axioms. So the groundwork consists of making sure that this yet-to-be-constructed θ exists and does what it is supposed to do. In this chapter we will specify our language and N, a set of nonlogical axioms. The axioms of N will be true sentences in $\mathfrak{N}$. We will show that N, although very weak, is strong enough to prove some crucial results. We will then show that our language is rich enough to express several ideas that will be crucial in the construction of θ.

In Chapter 5 we will prove Gödel's Self-Reference Lemma and use that lemma to construct the sentence θ. We shall then state and prove the First Incompleteness Theorem, that there can be no decidable, consistent, complete set of axioms for $\mathfrak{N}$. We then will finish the book with a discussion of Gödel's Second Incompleteness Theorem, which shows that no reasonably strong set of axioms can ever hope to prove its own consistency.

4.2 The Language, the Structure, and the Axioms of N

We work in the language of number theory

$$\mathcal{L}_{NT} = \{0, S, +, \cdot, E, <\},$$

and we will continue to work in this language for the next two chapters. $\mathfrak{N}$ is the standard model of the natural numbers,

$$\mathfrak{N} = \langle \mathbb{N}, 0, S, +, \cdot, E, < \rangle,$$

where the functions and relations are the usual functions and relations that you have known since you were knee high to a grasshopper. E is exponentiation, which will usually be written x^y rather than Exy or xEy.

We will now establish a set of nonlogical axioms, N. You will notice that the axioms are clearly sentences that are true in the standard structure, and thus if T is *any* set of axioms such that $T \vdash \sigma$ for all σ such that $\mathfrak{N} \models \sigma$, then $T \vdash N$. So, as we prove that several sorts of formulas are derivable from N, remember that those same formulas are also derivable from any set of axioms that has any hope of providing an axiomatization of the natural numbers.

The axiom system N was introduced in Example 2.8.3 and is reproduced on the next page. These eleven axioms establish some of the basic facts about the successor function, addition, multiplication, exponentiation, and the $<$ ordering on the natural numbers.

Chaff: To be honest, the symbol E and the axioms about exponentiation are not needed here. It is possible to do everything that we do in the next couple of chapters by defining exponentiation in terms of multiplication, and introducing E as an abbreviation in the language. This has the advantage of showing a little more explicitly how little you need to prove the incompleteness theorems, but adds some complications to the exposition. I have decided to introduce exponentiation explicitly and add a couple of axioms, which will hopefully allow us to move a little more cleanly through the proofs of our theorems.

The Axioms of N

1. $(\forall x)\neg Sx = 0$.
2. $(\forall x)(\forall y)\big[Sx = Sy \rightarrow x = y\big]$.
3. $(\forall x)x + 0 = x$.
4. $(\forall x)(\forall y)x + Sy = S(x + y)$.
5. $(\forall x)x \cdot 0 = 0$.
6. $(\forall x)(\forall y)x \cdot Sy = (x \cdot y) + x$.
7. $(\forall x)xE0 = S0$.
8. $(\forall x)(\forall y)xE(Sy) = (xEy) \cdot x$.
9. $(\forall x)\neg x < 0$.
10. $(\forall x)(\forall y)\big[x < Sy \leftrightarrow (x < y \vee x = y)\big]$.
11. $(\forall x)(\forall y)\big[(x < y) \vee (x = y) \vee (y < x)\big]$.

4.2.1 Exercises

1. You have already seen that N is not strong enough to prove the commutative law of addition (Exercise 8 in Section 2.8.1). Use this to show that N is not complete by showing that

$$N \nvdash (\forall x)(\forall y)x + y = y + x$$

and

$$N \nvdash \neg\big[(\forall x)(\forall y)x + y = y + x\big].$$

2. Suppose that Σ provides an axiomatization of $Th(\mathfrak{N})$. Suppose σ is a formula such that $N \vdash \sigma$. Show that $\Sigma \vdash \sigma$.

3. Suppose that $\mathfrak{A}$ is a nonstandard model of arithmetic. If $Th(\mathfrak{A})$ is the collection of sentences that are true in $\mathfrak{A}$, is $Th(\mathfrak{A})$ complete? Does $Th(\mathfrak{A})$ provide an axiomatization of $\mathfrak{N}$? Of $\mathfrak{A}$?

4.3 Recursive Sets and Recursive Functions

For the sake of discussion, suppose that we let $f(x) = x^2$. It will not surprise you to find out that it is the case that $f(4) = 16$, so I would like to write $\mathfrak{N} \models f(4) = 16$. Unfortunately, we are not allowed to do this, since the symbol f, not to mention 4 and 16, are not part of the language.

What we can do, however, is to represent the function f by a formula in $\mathcal{L}_{NT}$. To be specific, suppose that $\phi(x, y)$ is

$$y = ExSS0.$$

Then, if we allow ourselves once again to use the abbreviation $\overline{a}$ for the $\mathcal{L}_{NT}$-term $\underbrace{SSS \cdots S}_{aS\text{'s}} 0$, we can assert that

$$\mathfrak{N} \models \phi(\overline{4}, \overline{16})$$

which is the same thing as

$$\mathfrak{N} \models = SSSSSSSSSSSSSSSS0ESSSS0SS0.$$

(Boy, aren't you glad we don't use the official language very often?) Anyway, the situation is even better than this, for $\phi(\overline{4}, \overline{16})$ is derivable from N rather than just true in $\mathfrak{N}$. In fact, if you look back at Lemma 2.8.4, you probably won't have any trouble believing the following statements:

- $N \vdash \phi(\overline{4}, \overline{16})$
- $N \vdash \neg\phi(\overline{4}, \overline{17})$
- $N \vdash \neg\phi(\overline{1}, \overline{714})$

In fact, this formula ϕ is such, and N is such, that if a is any natural number and $b = f(a)$, then

$$N \vdash \forall y \big[\phi(\overline{a}, y) \leftrightarrow y = \overline{b}\big].$$

We will say that the formula ϕ represents the function f in the theory N.

Definition 4.3.1. A set $A \subseteq \mathbb{N}^k$ is said to be **recursive** if there is an $\mathcal{L}_{NT}$-formula $\phi(\underset{\sim}{x})$ such that

$$\begin{aligned} &\forall \underset{\sim}{a} \in A \quad N \vdash \phi(\overline{\underset{\sim}{a}}) \\ &\forall \underset{\sim}{b} \notin A \quad N \vdash \neg\phi(\overline{\underset{\sim}{b}}). \end{aligned}$$

In this case we will say that the formula ϕ **represents** the set A.

> *Chaff:* A bit of notation has slipped in here. Rather than writing $x_1, x_2, \ldots, x_k$ over and over and over again over the next few sections, we will abbreviate this as $\underset{\sim}{x}$. Similarly, $\overline{\underset{\sim}{x}}$ is shorthand for $\overline{x_1}, \overline{x_2}, \ldots, \overline{x_k}$. If you want, you can just assume that there is only one x—it won't make any difference to the exposition.

We will also spend a fair bit of time talking about recursive functions. Since a function $f : \mathbb{N}^k \to \mathbb{N}$ can be seen as a subset of $\mathbb{N}^{k+1}$, we can just think about the function as a set when proving it to be recursive and use Definition 4.3.1. In practice, we can do a little bit better:

If $f : \mathbb{N}^k \to \mathbb{N}$ is a recursive function, we can assume that f is represented by a formula $\phi(\underset{\sim}{x}, y)$ such that for every $\underset{\sim}{a} \in \mathbb{N}^k$,

$$N \vdash (\forall y)\left[\phi(\overline{\underset{\sim}{a}}, y) \leftrightarrow y = \overline{f(\underset{\sim}{a})}\right].$$

A proof of this is outlined in Exercise 7.

> *Chaff:*
>
> What's in a name? that which we call a rose
> By any other name would smell as sweet;
> So Romeo would, were he not Romeo call'd,
> Retain that dear perfection which he owes
> Without that title.
>
> —*Romeo and Juliet,* Act II, Scene ii
>
> In computer science courses, and in many mathematical logic texts, a different approach is taken when recursive sets and recursive functions are introduced. Starting with certain initial functions, the idea of recursion, and an object called the μ-operator, a collection of functions

is defined such that each function in the collection is effectively calculable. This is called the collection of recursive functions, which lead to something called *recursive sets,* whereas the sets that we have just christened as recursive are called *representable sets.* Then these texts prove that the collection of representable sets is the same as the collection of recursive sets. Thus we all end up at the same place, with nicely defined collection of sets and functions, that we call recursive. So if you are confused by the different definitions, just remember that they all define the same concept, and remember that the objects that are recursive (or representable) are (in some sense) the simple ones, the ones where membership can be proved in N.

To be fair, it is not quite as simple as Juliet makes it out to be. (It never is, is it?) The path that we have taken to recursive sets is clean and direct but emphasizes the deductions over the functions. The approach through initial functions stresses the fact that everything that we discuss can be calculated, and that viewpoint gives a natural tie between the logic that we have been discussing and its applications to computer science. For more on this connection, see Section 4.4.

Definition 4.3.2. We will say that a set $A \subseteq \mathbb{N}^k$ is **definable** if there is a formula $\phi(\underset{\sim}{x})$ such that

$$\begin{array}{ll} \forall \underset{\sim}{a} \in A & \mathfrak{N} \models \phi(\overline{\underset{\sim}{a}}) \\ \forall \underset{\sim}{b} \notin A & \mathfrak{N} \models \neg\phi(\overline{\underset{\sim}{b}}). \end{array}$$

In this case, we will say that ϕ **defines** the set A.

Chaff: It is very important to notice the difference between saying that ϕ represents A and ϕ defines A, which is the same as the difference between $N \vdash$ and $\mathfrak{N} \models$. Notice that any recursive set *must* be definable and is defined by any formula that represents it. The converse, however, is not automatic. In fact, the converse is not true. But we're getting ahead of ourselves.

We have mentioned several times that the axiom system N is relatively weak. We will show in this section that N is strong enough to prove some of the $\mathcal{L}_{NT}$ formulas that are true in $\mathfrak{N}$, namely the class of true Σ-sentences. And this will allow us to show that if a set A has a relatively simple *definition*, then the set A will be *recursive.*

Definition 4.3.3. If x is a variable that does not occur in the term t, let us agree to use the following abbreviations:

$$(\forall x < t)\phi \text{ means } \forall x(x < t \rightarrow \phi)$$

$$(\forall x \leq t)\phi \text{ means } \forall x((x < t \vee x = t) \rightarrow \phi)$$

$$(\exists x < t)\phi \text{ means } \exists x(x < t \wedge \phi)$$

$$(\exists x \leq t)\phi \text{ means } \exists x((x < t \vee x = t) \wedge \phi).$$

These abbreviations are called **bounded quantifiers**.

Definition 4.3.4. The collection of **Σ-formulas** is the smallest set of $\mathcal{L}_{NT}$-formulas that:

1. Contains all atomic formulas.
2. Contains all negations of atomic formulas.
3. Is closed under the connectives $\wedge$ and $\vee$.
4. Is closed under bounded quantifiers and the quantifier $\exists$.

The collection of **Π-formulas** is the smallest set of $\mathcal{L}_{NT}$-formulas that:

1. Contains all atomic formulas.
2. Contains all negations of atomic formulas.
3. Is closed under the connectives $\wedge$ and $\vee$.
4. Is closed under bounded quantifiers and the quantifier $\forall$.

The collection of **Δ-formulas** is the intersection of the collection of Σ-formulas with the set of Π-formulas.

Notice that a Δ-formula is a formula in which all of the quantifiers are bounded. The formula can have $\forall x < y$ and $\exists u \leq v$ but will not have any unbounded $\forall$'s or $\exists$'s.

Example 4.3.5. Here is a perfectly nice example of a Δ-formula ϕ: $(\forall x < t)(x = 0)$. Notice that the denial of ϕ is *not* a Δ-formula, as $\neg(\forall x < t)(x = 0)$ is neither a Σ- nor a Π-formula. But a chain of logical equivalences shows us that $\neg\phi$ is equivalent to a Δ-formula if we just push the negation sign inside the quantifier:

$$\neg\phi$$
$$\neg(\forall x < t)(x = 0)$$
$$(\exists x < t)\neg(x = 0).$$

Similarly, we can show that any propositional combination (using $\wedge, \vee, \neg, \rightarrow, \leftrightarrow$) of Δ-formulas is equivalent to a Δ-formula. We will use this fact approximately 215,342 times in the remainder of this book.

A Σ-sentence is, of course, a Σ-formula that is also a sentence. We will be particularly interested in Σ-formulas and Δ-formulas, for we will show if ϕ is a Σ-sentence and $\mathfrak{N} \models \phi$, then the axiom set N is strong enough to provide a deduction of ϕ. Since every Δ-sentence is also a Σ-sentence, any Δ-sentence that is true in $\mathfrak{N}$ is also provable from N.

The first lemma that we will prove shows that N is strong enough to prove that $1 + 1 = 2$. Actually, we already know this since it was proved back in Lemma 2.8.4. We now expand that result and show that if t is any variable-free term, then N proves that t is equal to what it is supposed to be equal to.

Recall that if t is a term, then $t^{\mathfrak{N}}$ is the interpretation of that term in the structure $\mathfrak{N}$. For example, suppose that t is the term $ESSS0SS0$, also known as $SSS0^{SS0}$. Then $t^{\mathfrak{N}}$ would be the number 9, and $\overline{t^{\mathfrak{N}}}$ would be the term $SSSSSSSSS0$. So when this lemma says that N proves $t = \overline{t^{\mathfrak{N}}}$, you should think that N proves $SSS0^{SS0} = SSSSSSSSS0$, which is the same as saying that $N \vdash \overline{3}^{\overline{2}} = \overline{9}$.

Lemma 4.3.6. *For each variable-free term t, $N \vdash t = \overline{t^{\mathfrak{N}}}$.*

Proof: We proceed by induction on the complexity of the term t. If t is the term 0, then $t^{\mathfrak{N}}$ is the natural number 0, and $\overline{t^{\mathfrak{N}}}$ is the term 0. Thus we have to prove that $N \vdash 0 = 0$, which is an immediate consequence of our logical axioms.

If t is $S(u)$, where u is a variable-free term, then the term $\overline{t^{\mathfrak{N}}}$ is identical to the term $S(\overline{u^{\mathfrak{N}}})$. Also, $N \vdash u = \overline{u^{\mathfrak{N}}}$, by the inductive hypothesis, and thus $N \vdash Su = S(\overline{u^{\mathfrak{N}}})$, thanks to the equality axiom (E2). Putting all of this together, we get that $N \vdash t = Su = S(\overline{u^{\mathfrak{N}}}) = \overline{t^{\mathfrak{N}}}$, as needed.

If t is $u+v$, we recall that Lemma 2.8.4 proved that $N \vdash \overline{u^{\mathfrak{N}}}+\overline{v^{\mathfrak{N}}} = \overline{u^{\mathfrak{N}} + v^{\mathfrak{N}}}$. But then $N \vdash t = u+v = \overline{u^{\mathfrak{N}}}+\overline{v^{\mathfrak{N}}} = \overline{u^{\mathfrak{N}} + v^{\mathfrak{N}}} = \overline{t^{\mathfrak{N}}}$, which is what we needed to show. The arguments for terms of the form $u \cdot v$ or u^v are similar, so the proof is complete. ■

The next lemma and its corollary will be used in our proof that true Σ-sentences are provable from N.

Lemma 4.3.7 (Rosser's Lemma). *If a is a natural number,*

$$N \vdash (\forall x < \overline{a})\big[x = \overline{0} \lor x = \overline{1} \lor \cdots \lor x = \overline{a-1}\big].$$

Proof: We use induction on a. If $a = 0$, it suffices to prove that $N \vdash \forall x[x < 0 \rightarrow \perp]$. By Axiom 9 of N, we know that $N \vdash \neg(x < 0)$, so $N \vdash (x < 0) \rightarrow \perp$, as needed.

For the inductive step, suppose that $a = b + 1$. We must show that

$$N \vdash \forall x\big[x < \overline{b+1} \rightarrow x = \overline{0} \lor \cdots \lor x = \overline{b}\big].$$

Since $\overline{b+1}$ and $S\overline{b}$ are identical, it suffices to show that

$$N \vdash \forall x\big[x < \overline{Sb} \rightarrow x = \overline{0} \lor \cdots \lor x = \overline{b}\big].$$

By Axiom 10, we know that $N \vdash x < S\overline{b} \rightarrow (x < \overline{b} \lor x = \overline{b})$, and then by the inductive hypothesis, we are finished. ■

Corollary 4.3.8. *If a is a natural number, then*

$$N \vdash \Big[\big[(\forall x < \overline{a})\phi(x)\big] \leftrightarrow \big[\phi(\overline{0}) \land \phi(\overline{1}) \land \cdots \land \phi(\overline{a-1})\big]\Big].$$

Proof: Exercise 9. ■

Now we come to the major result of this section, that our axiom system is strong enough to prove all true Σ-sentences.

Proposition 4.3.9. *If $\phi(\underset{\sim}{x})$ is a Σ-formula with free variables $\underset{\sim}{x}$, if $\underset{\sim}{t}$ are variable-free terms, and if $\mathfrak{N} \models \phi(\underset{\sim}{t})$, then $N \vdash \phi(\underset{\sim}{t})$.*

Proof: This is a proof by induction on the complexity of the formula ϕ.

1. If ϕ is atomic, say for example that ϕ is $x < y$ and terms t and u are such that $\mathfrak{N} \models t < u$. Then $t^{\mathfrak{N}} < u^{\mathfrak{N}}$, so by Lemma 2.8.4, $N \vdash \overline{t^{\mathfrak{N}}} < \overline{u^{\mathfrak{N}}}$. But we also know $N \vdash t = \overline{t^{\mathfrak{N}}}$ and $N \vdash u = \overline{u^{\mathfrak{N}}}$, by Lemma 4.3.6, so $N \vdash t < u$, as needed.

2. Negations of atomic formulas are handled in the same manner.

3. If ϕ is $\alpha \vee \beta$ or $\alpha \wedge \beta$, the argument is left to the Exercises.

4. Suppose that $\mathfrak{N} \models \exists x \psi(x)$, where we assume that ψ has only one free variable for simplicity. Then there is a natural number a such that $\mathfrak{N} \models \psi(\overline{a})$, and thus $N \vdash \psi(\overline{a})$ by the inductive hypothesis. But then our second quantifier axiom tells us, as $\overline{a}$ is substitutable for x in ψ, that $N \vdash \exists x \psi$, as needed.

5. Now if $\mathfrak{N} \models (\forall x < u)\psi(x)$, we know by the inductive hypothesis that

$$N \vdash \left[\psi(\overline{0}) \wedge \psi(\overline{1}) \wedge \cdots \wedge \psi(\overline{u^{\mathfrak{N}} - 1})\right].$$

But then by Corollary 4.3.8,

$$N \vdash (\forall x < \overline{u^{\mathfrak{N}}})\psi(x).$$

Thus, since $N \vdash \overline{u^{\mathfrak{N}}} = u$, $N \vdash (\forall x < u)\psi(x)$, as needed.

Thus if $\mathfrak{N} \models \phi(\underset{\sim}{t})$, then $N \vdash \phi(\underset{\sim}{t})$. ■

We will say that ϕ is **provable** (from N) if $N \vdash \phi$. And we shall say that ϕ is **refutable** if $N \vdash \neg\phi$.

Suppose that ϕ is a Δ-sentence. If $\mathfrak{N} \models \phi$, since we know that ϕ is also a Σ-sentence, Proposition 4.3.9 shows that $N \vdash \phi$. But suppose that ϕ is false; that is, suppose that $\mathfrak{N} \not\models \phi$. Then $\mathfrak{N} \models \neg\phi$, and $\neg\phi$ is equivalent to a Δ-sentence. Thus by the same argument as above, $N \vdash \neg\phi$. So we have proved the following:

Proposition 4.3.10. *If $\phi(\underset{\sim}{x})$ is a Δ-formula with free variables $\underset{\sim}{x}$, if $\underset{\sim}{t}$ are variable-free terms, and if $\mathfrak{N} \models \phi(\underset{\sim}{t})$, then $N \vdash \phi(\underset{\sim}{t})$. If, on the other hand, $\mathfrak{N} \models \neg\phi(\underset{\sim}{t})$, then $N \vdash \neg\phi(\underset{\sim}{t})$.*

Corollary 4.3.11. *Suppose that $A \subseteq \mathbb{N}^k$ is defined by a Δ-formula $\phi(\underset{\sim}{x})$. Then A is recursive.*

Proof: This is immediate from Proposition 4.3.10 and Definition 4.3.1. ■

Example 4.3.12. Suppose that we look at the even numbers. You might want to define this set by the $\mathcal{L}_{NT}$-formula

$$\phi(x) \text{ is: } (\exists y)(x = y + y).$$

But we can, in fact, do even better than this. We can define the set of evens by a Δ-formula

Even(x) is: $(\exists y \leq x)(x = y + y)$.

So now we have a Δ-definition of Even, the set of even numbers. (We will try to be consistent and use Small Capitals when referring to a set of numbers and *Italics* when referring to the $\mathcal{L}_{NT}$-formula that defines that set.) So by Corollary 4.3.11, we see that the set of even numbers is a recursive subset of the natural numbers.

Over the next few sections we will be doing a lot of this. We will look at a set of numbers and prove that it is recursive by producing a Δ-definition of the set. In many cases, the tricks that we will use to produce the bounds on the quantifiers will be quite impressive.

> *Chaff:* For the rest of this chapter you will see lots of formulas with boxes around them. The idea is that every time we introduce a Δ-definition of a set of numbers, there will be a box around it to set it off. I have also gathered three tables of some of the useful Δ-definitions on the endpapers of this book for (relatively) easy reference.

Example 4.3.13. Take a minute and write $\boxed{Prime(x)}$, a Δ-definition of Prime, the set of prime numbers. Once you have done that, here is a definition of the set of prime pairs, the set of pairs of numbers x and y such that both x and y are prime, and y is the next prime after x:

> $Primepair(x, y)$ is:
> $Prime(x) \land Prime(y) \land (x < y) \land \left[(\forall z < y)(Prime(z) \rightarrow z \leq x) \right]$.

Notice that *Primepair* has two free variables, as PRIMEPAIR $\subseteq \mathbb{N}^2$, while your formula *Prime* has exactly one free variable. Also notice that all of the quantifiers in each definition are bounded, so we know the definitions are Δ-definitions.

> *Chaff:* I also hope that you noticed that in the definition of *Primepair* I used your formula *Prime*, and I did not try to insert your entire formula every time I needed it—I just wrote $Prime(x)$ or $Prime(y)$. As you work out the many definitions to follow, it will be essential for you to do the same. Freely use previously defined formulas and plug them in by using their names. To do otherwise is to doom yourself to unending streams of unintelligible symbols. This stuff gets quite dense enough as it is. You do not need to make things any harder than they are.

4.3.1 Exercises

1. Show that the set $\{17\}$ is recursive by finding a Δ-formula that defines the set. Can you come up with a (probably silly) non-Δ formula that defines the same set?

2. Suppose that $A \subseteq \mathbb{N}$ is recursive and represented by the formula $\phi(x)$. Suppose also that $B \subseteq \mathbb{N}$ is recursive and represented by $\psi(x)$. Show that the following sets are also recursive, and find a formula that represents each:

(a) $A \cup B$
(b) $A \cap B$
(c) The complement of A, $\{x \in \mathbb{N} \mid x \notin A\}$

3. Show that every finite subset of the natural numbers is recursive and that every subset of $\mathbb{N}$ whose complement is finite is also recursive.

4. Let $p(x)$ be a polynomial with nonnegative integer coefficients. Show that the set $\{a \in \mathbb{N} \mid p(a) = 0\}$ is recursive.

5. Write a Δ-definition for the set DIVIDES. So you must come up with a formula with two free variables, $\boxed{Divides(x, y)}$, such that $\mathfrak{N} \models Divides(\overline{a}, \overline{b})$ if and only if a is a factor of b.

6. Show that the set $\{1, 2, 4, 8, 16, \dots\}$ of powers of 2 is recursive.

7. Suppose that you know that $\alpha(x, y)$ represents the function f. Show that $\phi(x, y) = \alpha(x, y) \land (\forall z < y)(\neg\alpha(x, z))$ also represents f, and furthermore, for each $a \in \mathbb{N}$,

$$N \vdash (\forall y)(\phi(\overline{a}, y) \leftrightarrow y = \overline{f(a)}).$$

[*Suggestion:* After you show that ϕ represents f, the second part is equivalent to showing $N \vdash \phi(\overline{a}, \overline{f(a)})$, which is pretty trivial, and then proving that

$$N \vdash \left[(\alpha(\overline{a}, y) \land (\forall z < y)(\neg\alpha(x, z))) \rightarrow y = \overline{f(a)}\right].$$

So, take as hypotheses N, $\alpha(\overline{a}, y)$, and $(\forall z < y)(\neg\alpha(x, z))$ and show that there is a deduction of both $\neg\left[\overline{f(a)} < y\right]$ and $\neg\left[y < \overline{f(a)}\right]$. Then the last of the axioms of N will give you what you need. For the details, see [Enderton 72, Theorem 33K].]

8. In the last inductive step of the proof of Lemma 4.3.6, the use of the inductive hypothesis is rather hidden. Please expose the use of the inductive hypothesis and write out that step of the proof more completely. Finish the cases for multiplication and exponentiation.

9. Prove Corollary 4.3.8.

10. Fill in the details of the steps omitted in the inductive proof of Proposition 4.3.9. In the last two cases, how does the argument change if there are more free variables? If, for example, instead of ϕ being of the form $\exists x \psi(x)$, ϕ is of the form $\exists x \psi(x, y)$, does that change the proof?

4.4 Recursive Sets and Computer Programs

In this section we shall investigate the relationship between recursive sets and computer programs. Our discussion will be rather informal and will rely on your intuition about computers and computation.

One of the reasons that we must be rather informal when discussing computation is that the idea of a computation is rather vague. In the mid-1930s many mathematicians developed theoretical constructs that tried to capture the idea of a computable function. Kurt Gödel's recursive functions, the Turing Machines of Alan Turing, and Alonzo Church's λ-calculus are three of the best-known models of computability.

One of the reasons that mathematicians accept these formal constructs as accurately modeling the intuitive notion of computability is that all of the formal analogs of computation that have been proposed have been proved to be equivalent. Thus it is known that a function is Turing computable if and only if it is general recursive if and only if it is λ-computable. It is also known that these formal notions are equivalent to the idea of a function being computable on a computer, where we will say that a function f is computable on an idealized computer if there is a computer program P such that if the program P is run with input n, the program will cause the computer to output $f(n)$ and halt.

Thus the situation is like this: On one hand, we have an intuitive idea of what it means for a function to be effectively calculable. On the other hand, we have a slew of formal models of computation, each of which is known to be equivalent to all of the others:

Intuitive Notion	Formal Models
Computable function	Recursive function λ-Computable function Turing-computable function Computer-computable function $\vdots$

So why do I say that the idea of a computation is vague? Although all of the *current* definitions are equivalent, for all we know there might be a new definition of computation that you will come up with tonight over a beer. That new definition will be intuitively correct, in the sense that people who hear your definition agree that it is the "right" definition of what it means for a person to compute something, but your definition may well *not* be equivalent to the current definitions. This will be an earthshaking development and will give logicians and computer scientists plenty to think about for years to come.

You will go down in history as a brilliant person with great insight into the workings of the human mind!

You will win lots of awards and be rich and famous!

All right, I admit it. Not rich. Just famous.

OK. Maybe not famous. But at least well known in logic and computer science circles.

But until you have that beer we will have to go with the current situation, where we have several equivalent definitions that seem to fit our current understanding of the word *computation*. So our idea of what constitutes a computation is imprecise, even though there is precision in the sense that lots of people have thought about what the definition ought to be, and every definition that has been proposed so far has been proved (precisely) to be equivalent to every other definition that has been proposed.

Church's Thesis is simply an expression of the belief that the formal models of computation accurately represent the intuitive idea of a calculable function. We will state the thesis in terms of recursiveness, as we have been working with recursive functions and recursive sets.

Church's Thesis. *A function is computable if and only if it is recursive.*

Now it is important to understand that Church's Thesis is a "thesis" as opposed to a "theorem" and that it will never be a theorem. As an attempt to link an intuitive notion (computability) and a formal notion (recursiveness) it is not the sort of thing that could *ever* be proved. Proofs require formal definitions, and if we write down a formal definition of computable function, we will have subverted the meaning of the thesis.

So is Church's Thesis true? We can say that all of the evidence to date seems to suggest that Church's Thesis is true, but I am afraid that is all the certainty that we can have on that point. We have over 60 years' experience since the statement of the thesis, and over 3000 years since we started computing functions, but that only counts as anecdotal evidence. Even so, most, if not all, of the mathematical community accepts the identification of "computable" with "recursive" and thus the community accepts Church's Thesis as an article of faith.

One of the areas of mathematical logic that has been very active has been the field of recursion theory. Roughly, recursion theory uses the formal notions of computability and examines sets and functions that satisfy these definitions. For example, one theorem states

Theorem 4.4.1. *A subset A of the natural numbers is recursive if there is an idealized computer and a computer program P such that:*

1. *If $a \in A$ and P is run with input a, then P will cause the computer to output* 1 *and halt.*
2. *If $a \notin A$ and P is run with input a, then P will cause the computer to output* 0 *and halt.*

(Equivalently, for the cognoscenti, A is recursive if and only if its characteristic function is computable by an idealized computer.)

The proof of this theorem is essentially verifying my claim that the formal notions of "recursive function" and "function that is calculable by an idealized computer" are equivalent.

Now the class of recursive subsets of $\mathbb{N}$ can be extended by looking at the collection of sets such that there is a program which will tell you if a number is an element of the set, but does not have to do anything at all if the number is not an element of the set.

Definition 4.4.2. A set $A \subseteq \mathbb{N}$ is said to **recursively enumerable**, or **r.e.**, if there is a computer program P such that if $a \in A$, program P returns 1 on input a, and if $a \notin A$, program P does not halt when given input a.

It is easy to prove the following:

Proposition 4.4.3. *A set A is recursive if and only if both A and $\mathbb{N} - A$ are recursively enumerable.*

Proof: Exercise 4. ■

The exercises provide a little more practice in working with recursively enumerable sets.

The study of recursive functions is an important area of mathematical logic, and emphasizes the tie between logic and computer science. Although we have not attempted to do so in this book, it is possible to develop an entire presentation of Gödel 's Incompleteness Theorem based on recursive functions. In this setting, the statement of the Incompleteness Theorem is the statement that the collection of sentences that are provable-from-Σ, where Σ is an extension of N that is decidable and true-in-$\mathfrak{N}$, is a recursively enumerable set that is not recursive. Thus there is a significant difference between the collection of recursive sets and the collection of r.e. sets. One text that emphasizes computability in its treatment of Gödel's Theorem is [Keisler and Robbin 96].

4.4.1 Exercises

1. (a) Show that $A \subseteq \mathbb{N}$ is recursively enumerable if and only if A is listable, where a set is **listable** if there is computer program L such that L prints out, in some order or another, the elements of A.

(b) Show $A \subseteq \mathbb{N}$ is recursive if and only if A is listable in increasing order.

2. Suppose that $A \subseteq \mathbb{N}$ is infinite and recursively enumerable and show that there is an infinite set $B \subseteq A$ such that B is recursive.

3. Show that $A \subseteq \mathbb{N}$ is recursively enumerable if and only if there is a Σ-formula $\phi(x)$ such that ϕ defines A.

4. Prove Proposition 4.4.3. [*Suggestion:* First assume that A is recursive. This direction is easy. For the other direction, the assumption guarantees the existence of two programs. Think about writing a new program that runs these two programs in tandem—first you run one program for a minute, then you run the second program for a minute. ...]

4.5 Coding—Naïvely

If you know a child of a certain age, you have undoubtedly run across coded messages of the form

$$1\ 14 \quad 1\ 16\ 16\ 12\ 5 \quad 1 \quad 4\ 1\ 25$$

where letters are coded by numbers associated with their place in the alphabet. If we ignore the spaces above, we can think of the phrase as being coded by a single number, and that number can have special properties. For example, the code above is not prime and it is divisible by exactly five distinct primes. If we like, we could say that the coding allows us to assert the same statements about the phrase that has been coded. For example, if we take the phrase

The code for this phrase is even

and coded it as a number, you might notice that the code ends in 4, so you might be tempted to say that the phrase was correct in what it asserts.

What we are doing here is representing English statements as numbers, and investigating the properties of the numbers. We can do the same thing with statements of $\mathcal{L}_{NT}$. For example, if we take the sentence

$$= 0S0,$$

we could perhaps code this sentence as the number

$$4964250291131250000000,$$

and then we can assert things about the number associated with the string. For example, the code for $= 0S0$ is in the set of numbers that are divisible by 10, and it is in the set of numbers that are larger than the national debt. What will make all of this interesting is we can ask if the code for $= 0S0$ is in the set of all codes of sentences that can be deduced from N. Then we will be asking if our sentence is a theorem of N. If it is, then we will know that N is inconsistent. If it is not, then N is consistent. Thus we will have reduced the question of consistency of N to a question about numbers and sets!

This is the insight that let Gödel to the Incompleteness Theorem. Given any reasonable set of axioms A, Gödel showed a way to code the phrase

This phrase is not a theorem of A

as a sentence of $\mathcal{L}_{NT}$ and prove that this sentence cannot be in the collection of sentences that are provable from A. So he found a sentence that was true, but not provable. We will, in this chapter and the next, do the same thing.

But first, we will have to establish our coding mechanism. In this section we will not develop our official coding, but rather, a simplified version to give you a taste of the things to come. Let me describe the system I used for the example above.

Symbol	Symbol Number	Symbol	Symbol Number
$\neg$	1	$+$	13
$\vee$	3	$\cdot$	15
$\forall$	5	E	17
$=$	7	$<$	19
0	9	(	21
S	11	)	23
		v_i	$2i$

Table 4.1: Symbol Numbers for $\mathcal{L}_{NT}$

I started by assigning symbol numbers to the symbols of $\mathcal{L}_{NT}$, as given in Table 4.1. Notice that the symbol numbers are only assigned for the official elements of the language, so if you need to use any abbreviations, such as $\rightarrow$ or $\exists$, you will have to write them out in terms of their definitions.

Then I had to figure out a way to code up sequences of symbols. The idea here is pretty simple, as we will just take the symbol numbers and use them as exponents for the prime numbers, in order. For example, if we look at the expression

$$= 0S0$$

the sequence of symbol numbers this generates is

$$\langle 7, 9, 11, 9 \rangle,$$

so the code for the sequence would be

$$2^7 3^9 5^{11} 7^9,$$

which is also known as

$$4964250291131250000000,$$

the example that we looked at earlier.

Notice that our coding is effective, in the sense that it is easy, given a number, to find its factorization and thus to find the string that is coded by the number.

> *Chaff:* The word *easy* is used here in its mathematical sense, not in its computer science sense. In reality it can take unbelievably long to factor many numbers, especially numbers of the size that we will discuss.

Now, the problem with all of this is not that you would find it difficult to recognize code numbers, or to decode a given number, or anything like that. Rather, what turns out to be tricky is to show that N, our collection of axioms, is strong enough to be able to *prove* true assertions about the numbers. For example, we would like N to be able to show that the term

$$\overline{4964250291131250000000}$$

represents a number that is the code for an $\mathcal{L}_{NT}$-sentence. The details of showing that N has this strength will occupy us for the next several sections.

4.5.1 Exercises

1. Decode the message that begins this section.

2. Code up the following $\mathcal{L}_{NT}$-formulas using the method described in this section.

(a) $= +000$

(b) $= Ev_1Sv_2 \cdot Ev_1v_2v_1$

(c) $(= 00 \lor (\neg < 00))$

(d) $(\exists v_2)(< v_20)$

3. Find the $\mathcal{L}_{NT}$-formula that is represented by the following numbers. A calculator (or a computer algebra system) will be helpful. Write your answer in a form that normal people can understand—normal people with some familiarity with first-order logic and mathematics, that is.

(a) 334714343040000000000000

(b) 8682913670726270976000000

(c) $2^{21}3^{5}5^{6}7^{23}11^{21}13^{7}17^{6}19^{9}23^{23}$

4.6 Coding Is Recursive

A basic part of our coding mechanism will be the ability to code finite sequences of numbers as a single number. A number c is going to be a code for a sequence of numbers $\langle k_1, k_2, \ldots, k_n\rangle$ if and only if

$c = 2^{k_1}3^{k_2}\cdots p_n^{k_n}$, where p_n is the nth prime number. We assume, for convenience, that none of the k_i are equal to 0.

We show now that N is strong enough to recognize code numbers. In other words, we want to establish

Proposition 4.6.1. *The collection of numbers that are codes for finite sequences is a recursive set.*

Proof: It is easy to write a Δ-definition for the set of code numbers:

$$\begin{array}{l} \mathit{Codenumber}(c) \text{ is:} \\ \qquad \mathit{Divides}(SS0, c) \wedge (\forall z < c)(\forall y < z) \\ \Big[\big(\mathit{Prime}(z) \wedge \mathit{Divides}(z, c) \wedge \mathit{Primepair}(y, z)\big) \rightarrow \mathit{Divides}(y, c)\Big]. \end{array}$$

Notice that *Codenumber*(c) is a formula with one free variable, c. If you look at it carefully, all the formula says is that c is divisible by 2 and if any prime divides c, so do all the earlier primes. Since the definition above is a Δ-definition, Corollary 4.3.11 tells us that the set CODENUMBER is a recursive set. ■

Since CODENUMBER is recursive and *Codenumber* is a Δ-formula, we now know (for example) that

$$N \vdash \mathit{Codenumber}(\overline{18})$$

and

$$N \vdash \neg \mathit{Codenumber}(\overline{45}).$$

Now, suppose that we wanted to take a code number, c, and decode it. To find the third element of the sequence of numbers coded by c, I need to find the exponent of the third prime number. Thus, for N to be able to prove statements about the sequence coded by a number, N will need to be able to recognize the function that takes i and assigns it to the ith prime number, p_i. Proving that this function p is recursive is our next major goal.

Definition 4.6.2. The function $p : \mathbb{N} \rightarrow \mathbb{N}$ is the function such that $p(0) = 1$, and for each positive natural number i, $p(i)$ is the ith prime number. We will often write p_i instead of $p(i)$.

Proposition 4.6.3. *The function p that enumerates the primes is a recursive function.*

Proof: We start by constructing a measure of the primes. A number a will be in the set YARDSTICK if and only if a is of the form $2^1 3^2 5^3 \cdots {p_i}^i$ for some i. So the first few elements of the set are $\{2, 18, 2250, 5402250, 870037764750, \ldots\}$.

Yardstick(a) is:

$$\begin{gathered} Codenumber(a) \land \\ Divides(SS0, a) \land \neg Divides(SSSS0, a) \land \\ (\forall y < a)(\forall z < a)(\forall k < a) \\ \Big[\big(Divides(z, a) \land Primepair(y, z)\big) \rightarrow \\ \big(Divides(y^k, a) \leftrightarrow Divides(z^{Sk}, a)\big)\Big]. \end{gathered}$$

If we unravel this, all we have is that 2 divides a, 4 does not divide a, and if z is a prime number such that z divides a, then the power of z that goes into a is one more than the power of the previous prime that goes into a.

Now it is relatively easy to provide a Δ-definition of the function p:

IthPrime(i, y) is:

$$\begin{gathered} Prime(y) \land \\ (\exists a \leq (y^i)^i)\big[Yardstick(a) \land Divides(y^i, a) \land \neg Divides(y^{Si}, a)\big]. \end{gathered}$$

Notice the tricky bound on the quantifier! Here is the thinking behind that bound: If y is, in fact, the ith prime, then here is an $a \in$ YARDSTICK that shows this fact: $a = 2^1 3^2 \cdots y^i$. But then certainly a is less than or equal to $\underbrace{y^i y^i \cdots y^i}_{i \text{ terms}}$, and so $a \leq (y^i)^i$. This bound is, of course, much larger than a (the lone exception being when $i = 1$ and $y = 2$), but we will only be interested in the existence of bounds, and will pay almost no attention to making the bounds precise in any sense. ■

> *Chaff:* There is a bit of tension over notation that needs to be mentioned here. Suppose that we wished to discuss the seventeenth prime number, which happens to be 59, and that y is supposed to be equal to 59. The obvious way to assert this would be to state that $y = p_{17}$, but we will tend to use the explicit $\mathcal{L}_{NT}$-formula $IthPrime(17, y)$. Our choice will give us a great increase in consistency, as our formulas become rather more complicated over the rest of this chapter. We will tend to write all of our functions in this consistent, if not exactly intuitive manner.

Now we can use the function *IthPrime* to find each element coded by a number:

$IthElement(e, i, c)$ is:

$$Codenumber(c) \land (\exists y < c)(IthPrime(i, y) \land Divides(y^e, c) \land \neg Divides(y^{Se}, c)).$$

So intuitively, $IthElement(e, i, c)$ is true if c is a code and e is the number at position i of the sequence coded by c.

The length of the sequence coded by c is also easily found:

$Length(c, l)$ is:

$$Codenumber(c) \land (\exists y < c)\Big[(IthPrime(l, y) \land Divides(y, c) \land (\forall z < c)[PrimePair(y, z) \rightarrow \neg Divides(z, c)])\Big].$$

All this says is that if the lth prime divides c and the $(l + 1)$st prime does not divide c, then the length of the sequence coded by c is l.

4.6.1 Exercise

1. Decide if the following statements are true or false as statements about the natural numbers. Justify your answers.

   (a) $(5, 13) \in$ IthPrime

(b) $(1200, 3) \in$ LENGTH

(c) $IthElement(\overline{1}, \overline{2}, \overline{3630})$ (Why are there those lines over the numbers?)

4.7 Gödel Numbering

We now change our focus from looking at functions and relations on the natural numbers, where it makes sense to talk about recursive sets, to functions mapping strings of $\mathcal{L}_{NT}$-symbols to $\mathbb{N}$. We will establish our coding system for formulas, associating to each $\mathcal{L}_{NT}$-formula ϕ its Gödel number, $\ulcorner \phi \urcorner$.

Definition 4.7.1. The function $\ulcorner\ \urcorner$, with domain the collection of finite strings of $\mathcal{L}_{NT}$-symbols and codomain $\mathbb{N}$, is defined as follows:

$$\ulcorner s \urcorner = \begin{cases} 2^1 3^{\ulcorner \alpha \urcorner} & \text{if } s \text{ is } (\neg\alpha)\text{, where } \alpha \text{ is an } \mathcal{L}_{NT}\text{-formula} \\ 2^3 3^{\ulcorner \alpha \urcorner} 5^{\ulcorner \beta \urcorner} & \text{if } s \text{ is } (\alpha \vee \beta)\text{, where } \alpha \text{ and } \beta \text{ are } \mathcal{L}_{NT}\text{-formulas} \\ 2^5 3^{\ulcorner v_i \urcorner} 5^{\ulcorner \alpha \urcorner} & \text{if } s \text{ is } (\forall v_i)(\alpha)\text{, where } \alpha \text{ is an } \mathcal{L}_{NT}\text{-formula} \\ 2^7 3^{\ulcorner t_1 \urcorner} 5^{\ulcorner t_2 \urcorner} & \text{if } s \text{ is } = t_1 t_2\text{, where } t_1 \text{ and } t_2 \text{ are terms} \\ 2^9 & \text{if } s \text{ is } 0 \\ 2^{11} 3^{\ulcorner t \urcorner} & \text{if } s \text{ is } St\text{, with } t \text{ a term} \\ 2^{13} 3^{\ulcorner t_1 \urcorner} 5^{\ulcorner t_2 \urcorner} & \text{if } s \text{ is } +t_1 t_2\text{, where } t_1 \text{ and } t_2 \text{ are terms} \\ 2^{15} 3^{\ulcorner t_1 \urcorner} 5^{\ulcorner t_2 \urcorner} & \text{if } s \text{ is } \cdot t_1 t_2\text{, where } t_1 \text{ and } t_2 \text{ are terms} \\ 2^{17} 3^{\ulcorner t_1 \urcorner} 5^{\ulcorner t_2 \urcorner} & \text{if } s \text{ is } E t_1 t_2\text{, where } t_1 \text{ and } t_2 \text{ are terms} \\ 2^{19} 3^{\ulcorner t_1 \urcorner} 5^{\ulcorner t_2 \urcorner} & \text{if } s \text{ is } < t_1 t_2\text{, where } t_1 \text{ and } t_2 \text{ are terms} \\ 2^{2i} & \text{if } s \text{ is the variable } v_i \\ 3 & \text{otherwise.} \end{cases}$$

Notice that each symbol is associated with its symbol number, as set out in Table 4.1.

Example 4.7.2. This is a neat function, but the numbers involved get really big, really fast. Suppose that we work out the Gödel number for the formula ϕ, where ϕ is $(\neg = 0S0)$.

Since ϕ is a denial, the definition tells us that

$$\ulcorner \phi \urcorner \text{ is } 2^1 3^{\ulcorner = 0S0 \urcorner}.$$

So we need to find $\ulcorner = 0S0 \urcorner$, and by the "equals" clause in the definition,

$$\ulcorner = 0S0 \urcorner \text{ is } 2^7 3^{\ulcorner 0 \urcorner} 5^{\ulcorner S0 \urcorner}.$$

But $\ulcorner 0 \urcorner = 2^9$, and $\ulcorner S0 \urcorner = 2^{11} 3^{\ulcorner 0 \urcorner} = 2^{11} 3^{(2^9)}$. Now we're getting somewhere. Plugging things back in, we get

$$\ulcorner = 0S0 \urcorner \text{ is } 2^7 3^{(2^9)} 5^{[2^{11} 3^{(2^9)}]}$$

so the Gödel number for $(\neg = 0S0)$ is

$$\ulcorner \phi \urcorner \text{ is } 2^1 3^{2^7 3^{(2^9)} 5^{[2^{11} 3^{(2^9)}]}}.$$

> *Chaff:* To get an idea about how large this number is, consider the fact that the exponent on the 5 is $\ulcorner S0 \urcorner = 2^{11} 3^{(2^9)}$, which is
>
> 395742204565276981359699327426961020014622
> 018797636168707779330639922518966590197760
> 367399522017881177908886103953123727465543
> 866395298629306327869004362537831640066738
> 737141407294716362902985484718968825205601
> 824599890015416030883743179500368711 68,
>
> approximately 10^{247}.

Now, let us play a bit with the Gödel number of ϕ:

$$\begin{aligned} \ulcorner \phi \urcorner &= 2^1 3^{2^7 3^{(2^9)} 5^{[10^{247}]}} \\ &> 3^{5^{[10^{247}]}}, \end{aligned}$$

so if we take common logarithms, we see that

$$\log(\ulcorner \phi \urcorner) > 5^{[10^{247}]} \log 3$$

and taking logarithms again,

$$\begin{aligned} \log(\log(\ulcorner \phi \urcorner)) &> 10^{247} \log 5 + \log(\log 3) \\ &> 10^{246}. \end{aligned}$$

Hmm. ... So this means that $\log(\ulcorner\phi\urcorner)$ is bigger than $10^{[10^{246}]}$. But the common logarithm of a number is (approximately) the number of digits that it takes to express the number in base 10 notation, so we have shown that it takes more than $10^{[10^{246}]}$ digits to write out the Gödel number of ϕ. If you remember that a google is 10^{100} and a googleplex is $10^{10^{100}}$, $\ulcorner\phi\urcorner$ is starting to look like a pretty big number, but it gets better!

To write out a string of $10^{[10^{246}]}$ characters (assuming a million characters per mile, or about 16 characters per inch) would require far more than $10^{[10^{245}]}$ miles, which is far more than $10^{[10^{244}]}$ light years.

Or, to look at it another way, if we assume that we can put about 132 lines of type on an $8\frac{1}{2}$- by 11-inch piece of paper (using both sides), that works out to about $10^{[10^{245}]}$ pieces of paper, and since a ream of paper (500 sheets) is about 2 inches thick, that gives a stack of paper more than $10^{[10^{244}]}$ light years high. Since the age of the universe is currently estimated to be in the tens of billions of years (on the order of 10^{10} years), if we assume that the universe is both Euclidean and spherical, the volume of the universe is less than 10^{40} cubic light years, rather less than the $10^{[10^{244}]}$ cubic light years we would need to store our stack of paper. In short, we don't win any prizes for being incredibly efficient with the coding that we have chosen. What we do win is ease of analysis. The fact that we have chosen to code recursively will make our proofs to come much easier to comprehend.

4.7.1 Exercises

1. Evaluate the Gödel number for each of the following.

(a) $(\forall v_3)(v_3 + 0 = v_4)$

(b) $SSSS0$

2. Find the formula or term that is coded by each of the following:

(a) $2^7 3^{\left[2^{13} 3^{\left(2^{17} 3^4 5^{1024}\right)} 5^{\left(2^{17} 3^{16} 5^{1024}\right)}\right]} 5^{\left[2^{17} 3^{64} 5^{1024}\right]}$

(b) $2^1 3^{\left(2^{19} 3^{2^9} 5^{2^{11} 3^{2^9}}\right)}$

(c) $2^5 3^{\left(2^2\right)} 5^{\left(2^7 3^4 5^4\right)}$

4.8 Gödel Numbers and N

Suppose I asked you if the number

$$
\begin{aligned}
a \quad = \quad & 98935551141319245339924831856740255003655504699409 04 \\
& 217694483265998062974164754944009184988050447029 4477 \\
& 221525988280931866385966598824657326581967251090 6344 \\
& 579100166846842853518236790907257463711797422063 0140 \\
& 04561499725038540077209357948750921779 20000
\end{aligned}
$$

was the Gödel number of an $\mathcal{L}_{NT}$-term. How would you go about finding out? A reasonable approach would be to factor a and try to decode. It turns out that $a = 2^{13} 3^{512} 5^4$, and since $512 = 2^9 = \ulcorner 0 \urcorner$ and $4 = 2^2 = \ulcorner v_1 \urcorner$, you know that a is the Gödel number of the term $+0v_1$. That was easy.

What makes this more interesting is that the above is so easy that N can *prove* that a is the Gödel number of a term. Establishing this fact is the goal of this section.

We will show how to construct certain Δ-formulas, for example the formula $Term(x)$, such that for every natural number a, $\mathfrak{N} \models Term(\overline{a})$ if and only if a is the Gödel number of a term. Since our formula will be a Δ-formula, this tells us that $N \vdash Term(\overline{a})$ if a is the Gödel number of a term, and $N \vdash \neg Term(\overline{a})$ if a is not the Gödel number of a term.

The problem is going to be in writing down the formula. As you look at the definition of the function $\ulcorner\ \urcorner$ in Definition 4.7.1, you can see that the definition is by recursion, and we will need a way to deal with recursive definitions within the constraints of Δ-definitions. We would like to be able to write something like

$$Term(x) \text{ is: } \cdots \vee (\exists y < x)\left[x = 2^{11} 3^y \wedge Term(y)\right] \vee \cdots ,$$

but this definition is clearly circular. A technical trick will get us past this point.

But we should start at the beginning. You know that the collection of $\mathcal{L}_{NT}$-terms is the closure of the set of variables and the collection of constant symbols under the function symbols. We will begin by showing that the collection of Gödel numbers of variables is recursive.

Lemma 4.8.1. *The set*

$$\text{VARIABLE} = \{a \in \mathbb{N} \mid a = \ulcorner v \urcorner \textit{ for some variable } v\}$$

is recursive.

Proof: It suffices to provide a Δ-definition for VARIABLE:

> *Variable*(x) is:
> $$(\exists y < x)(Even(y) \wedge 0 < y \wedge x = 2^y).$$

Notice that we use the fact that if $x = 2^y$, then $y < x$. It is easy to see that $\mathfrak{N} \models Variable(\overline{a})$ if and only if $a \in$ VARIABLE, so our formula shows that VARIABLE is a recursive set. ■

To motivate our development of the formula *Term*, consider the term t, where t is $+0Sv_1$. We are used to recognizing that this is a term by looking at it from the outside in: t is a term, as it is the sum of two terms. Now we need to start looking at t from the inside out: t is a term, as there is a sequence of terms, each of which is either a constant symbol, a variable, or constructed from earlier entries in the sequence by application of a function symbol of the appropriate arity. Here is a construction sequence for our term t:

$$\langle v_1, Sv_1, 0, +0Sv_1 \rangle.$$

From this construction sequence we can look at the associated sequence of Gödel numbers:

$$\langle 2^2, 2^{11}3^{2^2}, 2^9, 2^{13}3^{2^9}5^{(2^{11}3^{2^2})} \rangle,$$

and then we can code this sequence up as a single number, as in Section 4.6:

$$c = 2^{[2^2]}3^{[2^{11}3^{2^2}]}5^{[2^9]}7^{\left[2^{13}3^{2^9}5^{(2^{11}3^{2^2})}\right]}.$$

Now we can begin to see what our formula $Term(a)$ is going to look like. We will know that a is the Gödel number of a term if there is a number c that is the code for a construction sequence for a term, and the last number in that construction sequence is a. We begin by defining the collection of construction sequences:

Definition 4.8.2. A finite sequence of $\mathcal{L}_{NT}$-terms $\langle t_1, t_2, \ldots, t_l \rangle$ is called a **term construction sequence for** t_l if, for each i, $1 \leq i \leq l$, t_i is either a variable, the constant symbol 0, or is one of $t_j, St_j, +t_j t_k, \cdot t_j t_k$, or $Et_j t_k$, where $j < i$ and $k < i$.

Proposition 4.8.3. *The set*

$$\begin{aligned}&\textsc{TermConstructionSequence} = \\ &\qquad \{(c, a) \mid c \text{ codes a term construction sequence} \\ &\qquad\qquad \text{for the term with Gödel number } a\}\end{aligned}$$

is recursive.

Proof: Here is a Δ-definition for the set:

$$\begin{aligned}
&\mathit{TermConstructionSequence}(c, a) \text{ is:} \\
&\quad \mathit{CodeNumber}(c) \land \\
&\quad (\exists l < c)\Big[\mathit{Length}(c, l) \land \mathit{IthElement}(a, l, c) \land \\
&\quad (\forall e < c)(\forall i \leq l)\Big(\mathit{IthElement}(e, i, c) \rightarrow \\
&\quad \mathit{Variable}(e) \lor e = \overline{2^9} \lor \\
&\quad (\exists j < i)(\exists k < i)(\exists e_j < c)(\exists e_k < c) \\
&\quad (\mathit{IthElement}(e_j, j, c) \land \mathit{IthElement}(e_k, k, c) \land \\
&\quad [e = e_j \lor e = \overline{2^{11}} \cdot \overline{3}^{e_j} \lor \\
&\quad e = \overline{2^{13}} \cdot \overline{3}^{e_j} \cdot \overline{5}^{e_k} \lor \\
&\quad e = \overline{2^{15}} \cdot \overline{3}^{e_j} \cdot \overline{5}^{e_k} \lor e = \overline{2^{17}} \cdot \overline{3}^{e_j} \cdot \overline{5}^{e_k}])\Big)\Big].
\end{aligned}$$

This just says that $(c, a) \in$ TermConstructionSequence if and only if c is a code of length l, a is the last number of the sequence coded by c, and if e is an entry at position i of c, then e is either the Gödel number of a variable, the Gödel number of 0, a repeat of an earlier entry, or is the Gödel number that is the result of applying S, $+$, $\cdot$, or E to earlier entries in c. As all of the quantifiers are bounded, this is a Δ-definition, so the set TERMCONSTRUCTIONSEQUENCE is recursive. ■

Now it would seem that to define *Term*(a), all we would have to do is to say that a is the Gödel number of a term if there is a number c such that *TermConstructionSequence*(c, a). This is not quite enough, as the quantifier $\exists c$ is not bounded. In order to write down a Δ-definition of *Term*, we will have to get a handle on how large construction sequences have to be.

Lemma 4.8.4. *If t is an $\mathcal{L}_{NT}$-term and $\ulcorner t \urcorner = a$, then the number of symbols in t is less than a.*

Proof: The proof is by induction on the complexity of t. Just to give you an idea of how true the lemma is, consider the example of t, where t is $S0$. Then t has two symbols, while $\ulcorner t \urcorner = a = 2^{11}3^{512}$, which is just a little bigger than 2. ■

Lemma 4.8.5. *If t is a term, the length of the shortest construction sequence of t is less than or equal to the number of symbols in t.*

Proof: Again, use induction on the complexity of t. ■

These lemmas tell us that if a is the Gödel number of a term, then there is a construction sequence of that term whose length is less than a.

Lemma 4.8.6. *Suppose that t is an $\mathcal{L}_{NT}$-term and u is a subterm of t (in other words, u is a substring of t and u is also an $\mathcal{L}_{NT}$-term). Then $\ulcorner u \urcorner < \ulcorner t \urcorner$.*

Proof: Exercise 4. ■

Lemma 4.8.7. *If a is a natural number greater than or equal to 1, then $p_a \leq 2^{a^a}$, where p_a is the ath prime number.*

Proof: Exercise 5. ■

Now we have enough to give us our bound on the code for the shortest construction sequence for a term t with Gödel number $\ulcorner t \urcorner = a$. Any such construction sequence must look like

$$\langle t_1, t_2, \ldots, t_k = t \rangle,$$

where $k \leq a$ and each t_i is a subterm of t. But then the code for this construction sequence is

$$\begin{aligned} c &= 2^{\ulcorner t_1 \urcorner} 3^{\ulcorner t_2 \urcorner} \cdots p_k^{\ulcorner t \urcorner} \\ &\leq 2^a 3^a \cdots p_k^a \\ &\leq \underbrace{p_k^a p_k^a \cdots p_k^a}_{k \text{ terms}} \\ &\leq \underbrace{p_a^a p_a^a \cdots p_a^a}_{a \text{ terms}} \\ &= \left[\left(2^{a^a} \right)^a \right]^a \\ &= \left(2^{a^a} \right)^{a^2}, \end{aligned}$$

which gives us our needed bound.

We are finally at a position where we can give a Δ-definition of the collection of Gödel numbers of $\mathcal{L}_{NT}$-terms:

> $Term(a)$ is:
>
> $$\left(\exists c < \left(\overline{2}^{a^a} \right)^{a^{\overline{2}}} \right) TermConstructionSequence(c, a).$$

So the set

$$\text{TERM} = \{ a \in \mathbb{N} \mid a \text{ is the Gödel number of an } \mathcal{L}_{NT}\text{-term} \}$$

is a recursive set, and thus N has the strength to prove, for any number a, either $Term(\overline{a})$ or $\neg Term(\overline{a})$. In Exercise 6 we will ask you to show that the set FORMULA is also recursive.

4.8.1 Exercises

1. Assume that ϕ is a formula of $\mathcal{L}_{NT}$. Which of the following are also $\mathcal{L}_{NT}$-formulas? For the ones that are not formulas, why are they not formulas?

- $Term(\phi)$
- $Term(\ulcorner \phi \urcorner)$
- $Term(\overline{\ulcorner \phi \urcorner})$

2. Suppose that in the definition of *TermConstructionSequence*, you saw the following string:

$$\cdots \lor e = \overline{2^{11} \cdot 3^{e_j}} \lor \cdots$$

Would that be a part of a legal $\mathcal{L}_{NT}$-formula? How do you know? What if the string were

$$\cdots \lor e = \overline{2}^{\overline{11}} \cdot \overline{3}^{\overline{e_j}} \lor \cdots ?$$

3. Prove Lemma 4.8.4.

4. Prove Lemma 4.8.6.

5. Prove Lemma 4.8.7 by induction. For the inductive step, if you are trying to prove that $p_{n+1} \leq 2^{(n+1)^{n+1}}$, use the fact that p_{n+1} is less than or equal to the smallest prime factor of $(p_1 p_2 \cdots p_n) - 1$.

6. A proof similar to the proof that TERM is recursive will show that

$$\text{FORMULA} = \{a \in \mathbb{N} \mid a \text{ is the Gödel number of an } \mathcal{L}_{NT}\text{-formula}\}$$

is also recursive. Carefully supply the needed details and define the formula $\boxed{Formula(f)}$. You will probably have to define the formula *FormulaConstructionSequence*(c, f) and estimate the length of such sequences as part of your exposition.

4.9 NUM and SUB Are Recursive

In our proof of the Self-Reference Lemma in Section 5.2, we will have to be able to substitute the Gödel number of a formula into a formula. To do this it will be necessary to know that a couple of functions are recursive, and in this section we outline how to construct Δ-definitions of those functions. First we work with the function Num.

Recall that $\overline{a}$ is the numeral representing the number a. Thus, $\overline{2}$ is $SS0$. Since $SS0$ is an $\mathcal{L}_{NT}$-term, it has a Gödel number, in this case

$$\ulcorner SS0 \urcorner = 2^{11}3^{\ulcorner S0 \urcorner} = 2^{11}3^{2^{11}3^{\ulcorner 0 \urcorner}} = 2^{11}3^{2^{11}3^{(2^9)}}.$$

The function Num that we seek will map 2 to the Gödel number of its $\mathcal{L}_{NT}$-numeral, $2^{11}3^{2^{11}3^{(2^9)}}$. So $\text{Num}(a) = \ulcorner \overline{a} \urcorner$.

To write a Δ-definition $Num(a, y)$ we will start, once again, with a construction sequence, but this time we will construct the numeral $\overline{a}$. For $a = 2$, the construction sequence is

$$\langle 0, S0, SS0 \rangle,$$

which gives rise to the sequence of Gödel numbers

$$\langle 2^9, 2^{11}3^{2^9}, 2^{11}3^{2^{11}3^{(2^9)}} \rangle,$$

which is coded by the number

$$c = 2^{[2^9]}3^{[2^{11}3^{2^9}]}5^{[2^{11}3^{2^{11}3^{(2^9)}}]}.$$

Notice that the length of the construction sequence here is 3, and in general the construction sequence will have length $a + 1$ if we seek to code the construction of the Gödel number of the numeral associated with the number a. (That is a very long sentence, but it does make sense if you work through it carefully.)

You are asked in Exercise 1 to write down the formula

$$\boxed{NumConstructionSequence(c, a, y)}$$

as a Δ-formula. The idea is that

$$\mathfrak{N} \models NumConstructionSequence(c, a, y)$$

if and only if c is the code for a construction sequence of length $a + 1$ with last element $y = \ulcorner \overline{a} \urcorner$.

Now we would like to define the formula $Num(a, y)$ in such a way that $Num(a, y)$ is true if and only if y is $\ulcorner \overline{a} \urcorner$, and as in Section 4.8, the formula $(\exists c) NumConstructionSequence(c, a, y)$ does not work, as the quantifier is unbounded. So we must find a bound for c.

If $(c, a, y) \in$ NumConstructionSequence, then we know that c codes a construction sequence of length $a+1$ and c is of the form

$$c = 2^{\ulcorner t_1 \urcorner} 3^{\ulcorner t_2 \urcorner} \cdots p_{a+1}^{y},$$

where each t_i is a subterm of $\overline{a}$. By Lemma 4.8.6 and Lemma 4.8.7, we know that $\ulcorner t_i \urcorner \leq y$ and $p_{a+1} \leq 2^{(a+1)^{(a+1)}}$, so

$$c \leq \underbrace{2^y 3^y \ldots p_{a+1}^y}_{a+1 \text{ terms}} \leq (p_{a+1})^{(a+1)y} \leq \left(2^{(a+1)^{(a+1)}}\right)^{(a+1)y}.$$

This gives us our needed bound on c. Now we can define

> $Num(a, y)$ is:
> $$\left(\exists c < \left(\overline{2}^{(a+\overline{1})^{(a+\overline{1})}}\right)^{(a+\overline{1}) \cdot y}\right)$$
> $$NumConstructionSequence(c, a, y).$$

The next formulas that we need to discuss will deal with substitution. We will define two Δ-formulas:

1. $TermSub(u, x, t, y)$ will represent substitution of a term for a variable in a term (u_t^x).

2. $Sub(f, x, t, y)$ will represent substitution of a term for a variable in a formula (ϕ_t^x).

Specifically, we will show that $\mathfrak{N} \models TermSub(\ulcorner u \urcorner, \ulcorner x \urcorner, \ulcorner t \urcorner, y)$ if and only if y is the Gödel number of u_t^x, where it is assumed that u and t are terms and x is a variable. Similarly, $Sub(\ulcorner \phi \urcorner, \ulcorner x \urcorner, \ulcorner t \urcorner, y)$ will be true if and only if ϕ is a formula and $y = \ulcorner \phi_t^x \urcorner$.

We will develop *TermSub* carefully and outline the construction of *Sub*, leaving the details to the reader.

First, let us look at an example. Suppose that u is the term $+ \cdot 0S0x$. Then a construction sequence for u could look like

$$\langle 0, x, S0, \cdot 0S0, + \cdot 0S0x \rangle.$$

If t is the term $SS0$, then u_t^x is $+ \cdot 0S0SS0$, and we will find a construction sequence for this term in stages.

The first thing we will do is look at the sequence (no longer a construction sequence) that we obtain by replacing all of the x's in u's construction sequence with t's:

$$\langle 0, SS0, S0, \cdot 0S0, + \cdot 0S0SS0 \rangle.$$

This fails to be a construction sequence, as the second term of the sequence is illegal. However, if we precede this with the construction sequence for t, all will be well (recall that we are allowed to repeat elements in a construction sequence):

$$\langle 0, S0, SS0, 0, SS0, S0, \cdot 0S0, + \cdot 0S0SS0 \rangle.$$

So to get all of this to work, we have to do three things:

1. We must show how to change u's construction sequence by replacing the occurrences of x with t.

2. We must show how to put one construction sequence in front of another.

3. We must make sure that all of our quantifiers are bounded throughout, so that our final formula *TermSub* is a Δ-formula.

So, first we define a formula $TermReplace(c, u, d, x, t)$ such that if c is the code for a construction sequence of a term with Gödel number u, then d is the code of the sequence (probably not a construction sequence) that results from replacing each occurrence of the variable with Gödel number x by the term with Gödel number t. In the following definition, the idea is that e_i and a_i are the entries at position i of the sequence coded by c and d, respectively, and if we look at it line by line we see: c codes a construction sequence for the term coded by u, d codes a sequence; the lengths of the two sequences are the same; if e_i and a_i are the ith entries in sequence c and d, respectively, then e_i is x if and only if a_i is t; e_i is another variable if and only if a_i is the same variable; e_i codes 0 if and only if a_i also codes 0; e_i codes the successor of a previous e_j if and only if a_i codes the successor of the corresponding a_j; and so on.

$$
\begin{aligned}
&\mathit{TermReplace}(c,u,d,x,t) \text{ is:} \\
&\quad \mathit{TermConstructionSequence}(c,u) \land \mathit{Codenumber}(d) \land \\
&\quad (\exists l < c)\Big[\mathit{Length}(c,l) \land \mathit{Length}(d,l) \land \\
&\quad (\forall i \leq l)(\forall e_i < c)(\forall a_i < d) \\
&\quad \Big((\mathit{IthElement}(e_i,i,c) \land \mathit{IthElement}(a_i,i,d)) \rightarrow \\
&\quad [([(\mathit{Variable}(e_i) \land e_i = x) \leftrightarrow a_i = t] \land \\
&\quad (\mathit{Variable}(e_i) \land e_i \neq x) \leftrightarrow (\mathit{Variable}(a_i) \land a_i = e_i)) \land \\
&\quad (e_i = \overline{2}^{\overline{9}} \leftrightarrow a_i = \overline{2}^{\overline{9}}) \land \\
&\quad (\forall j < i)\Big[((\exists e_j < c)\mathit{IthElement}(e_j,j,c) \land e_i = \overline{2}^{\overline{11}} \cdot \overline{3}^{e_j}) \rightarrow \\
&\quad ((\exists a_j < d)\mathit{IthElement}(a_j,j,d) \land a_i = \overline{2}^{\overline{11}} \cdot \overline{3}^{a_j})\Big] \land \\
&\quad \vdots \\
&\quad \text{(Similar clauses for } +,\ \cdot, \text{ and } E.) \\
&\quad \vdots \\
&\quad]\Big)\Big].
\end{aligned}
$$

Now that we know how to destroy a construction sequence for u by replacing all occurrences of x with t, we have to be able to put a term construction sequence for t in front of our sequence to make it a construction sequence again. So suppose that d is the code for the sequence obtained by replacing x by t, and that b is the code for the term construction sequence for t. Essentially, we want a to code d appended to b. So the length of a is the sum of the lengths of b and d, and the first l elements of the a sequence should be the same as the b sequence, where the length of the b sequence is l, and the rest of the a sequence should match, entry by entry, the d sequence:

$$\begin{array}{l}
Append(b,d,a) \text{ is:} \\
\quad Codenumber(b) \land Codenumber(d) \land Codenumber(a) \land \\
\quad\quad (\exists l < b)(\exists m < d)\Big(Length(l,b) \land \\
\quad\quad Length(m,d) \land Length(l+m,a) \land \\
(\forall e < a)(\forall i \leq l)\big[IthElement(e,i,a) \leftrightarrow IthElement(e,i,b)\big] \land \\
\quad\quad (\forall e < a)(\forall j \leq m)\big[0 < j \rightarrow \\
\quad\quad (IthElement(e,l+j,a) \leftrightarrow IthElement(e,j,d))\big]\Big).
\end{array}$$

Now we are ready to define the formula $TermSub(u,x,t,y)$ that is supposed to be true if and only if y is the (Gödel number of the) term that results when you take the term u and replace x by t. A first attempt at a definition might be

$$\begin{array}{l}
(\exists a)(\exists b)(\exists d)(\exists c)\big[\,TermConstructionSequence(c,u) \land \\
\qquad TermConstructionSequence(b,t) \land \\
\qquad TermReplace(c,u,d,x,t) \land Append(b,d,a) \land \\
\qquad\qquad TermConstructionSequence(a,y)\big].
\end{array}$$

This is nice, but we need to bound all of the quantifiers. Bounding b and d is easy: They are smaller than a. As for c, we showed when we defined the formula $Term(a)$ on page 155 that the term coded by u must have a construction sequence coded by a number less than $\left(2^{u^u}\right)^{u^2}$. So all that is left is to find a bound on a. Notice that we can assume that b codes a sequence of length less than or equal to t, the last entry of b, and similarly, c codes a sequence of length less than or equal to u. Since the sequence coded by d has the same length as the sequence coded by c, and as a codes b's sequence followed by d's, the length of the sequence coded by a is less than or equal to $u + t$. So

$$a \leq 2^{e_1} 3^{e_2} \cdots p_{t+u}^{y}.$$

Now we can use the same trick that we used in the definition of *TermConstructionSequence*. As each entry of a can be assumed to be less than or equal to y, we get

$$\begin{aligned} a &\leq 2^y 3^y \dots p_{t+u}^y \\ &\leq [(p_{t+u})^y]^{t+u} \\ &\leq \left(\left[2^{t+u^{t+u}}\right]^y\right)^{t+u}. \end{aligned}$$

So our Δ-definition of TermSub is

> $TermSub(u, x, t, y)$ is:
>
> $$\left(\exists a < \left(\left[\overline{2}^{t+u^{t+u}}\right]^y\right)^{t+u}\right)(\exists b < a)(\exists d < a)\left(\exists c < \left(\overline{2}^{u^u}\right)^{u^{\overline{2}}}\right)$$
>
> $$\big[TermConstructionSequence(c, u) \land TermConstructionSequence(b, t) \land TermReplace(c, u, d, x, t) \land Append(b, d, a) \land TermConstructionSequence(a, y)\big].$$

Chaff: Oh, how I hate garbage cases. My claim in the paragraph preceding the definition of *TermSub* that each entry of a will be less than or equal to y might not be correct if $y = u_t^x = u$, so the substitution is vacuous. The reason for this is that the entries of a would include a construction sequence for t, which might be huge, while y might be relatively small. For example, we might have $u = 0$ and $t = v_{123456789}$ and $x = v_1$. In Exercise 3 you are invited to figure out the slight addition to the definition of *TermSub* that takes care of this.

Now we outline the construction of the formula $Sub(f, x, t, y)$, which is to be true if f is the Gödel number of a formula ϕ and y is $\ulcorner \phi_t^x \urcorner$. The idea is to take a formula construction sequence for ϕ and follow it by a copy of the same construction sequence where we systematically replace the occurrences of x with t's. You do have to be a little careful in your replacements, though. If you compare Definitions 1.8.1 and 1.8.2, you can see that the rules for replacing variables by terms are a little more complicated in the formula case than in the term case, particularly when you are substituting in a quantified formula. But the difficulties are not too bad.

So if b codes the construction sequence for ϕ, if d codes the sequence that you get after replacing the x's by t's, and if $Append(b, d, a)$ holds, then a will code up a construction sequence for ϕ_t^x. After you deal with the bounds, you will have a Δ-formula along the lines of

$Sub(f, x, t, y)$ is:

$$(\exists a < \text{Bound})(\exists b < a)(\exists d < a)$$
$$FormulaConstructionSequence(b, f) \wedge$$
$$FormulaReplace(b, f, d, x, t) \wedge Append(b, d, a) \wedge$$
$$FormulaConstructionSequence(a, y).$$

4.9.1 Exercises

1. Write the formula $NumConstructionSequence(c, a, y)$. Make sure that your formula is a Δ-formula with the variables c, a, and y free. [*Suggestion:* You might want to model your answer on the construction of *TermConstructionSequence*.]

2. Write out the "Similar clauses" in the definition of *TermReplace*.

3. Change the definition of *TermSub* to take care of the problem mentioned on page 162. [*Suggestion:* The problem occurs only if the substitution is vacuous. So there are two cases. Either u and y are different, in which case our definition is fine, or u and y are equal. What do you need to do then? So I suggest that your answer should be a disjunction, something like

$MyTermSub(u, x, t, y)$ is:

$$[(u \neq y) \wedge TermSub(u, x, t, y)] \vee$$
$$[(u = y) \wedge (\text{Something Brilliant})].$$

4. Write out the details of the formula *Sub*.

4.10 Definitions by Recursion Are Recursive

If you look at the definition of a term (Definition 1.3.1), the definition of formula (Definition 1.3.3), the definition of u_t^x (Definition 1.8.1), and the definition of ϕ_t^x (Definition 1.8.2), you will notice that all of these definitions were definitions "by recursion." For example, in the definition of a term, you see the phrase

> ... t is $ft_1t_2 \dots t_n$, where f in an n-ary function symbol of $\mathcal{L}$ and each of the t_i are terms of $\mathcal{L}$

so a term can have constituent parts that are themselves terms. In the last two sections we have used the device of construction sequences to show that the sets TERM, FORMULA, TERMSUB, and SUB are recursive sets. In this section we outline a proof that all such sets of strings, defined "by recursion," give rise to sets of Gödel numbers that are recursive. It will be clear from our exposition that a more general statement of our theorem could be proved, but what we present will be sufficient for our needs.

Definition 4.10.1. A string of symbols s from a first-order language $\mathcal{L}$ is called an **expression** if s is either a term of $\mathcal{L}$ or a formula of $\mathcal{L}$.

Theorem 4.10.2. *Suppose that we have a set of $\mathcal{L}_{NT}$-expressions, which we will call Set, defined as follows: An expression s is an element of Set if and only if:*

1. *s is an element of BaseCaseSet, or*

2. *There is an expression t, a proper substring of s, such that (t, s) is an element of ConstructionSet.*

If the sets of strings BaseCaseSet and ConstructionSet give rise to sets of Gödel numbers BASECASESET *and* CONSTRUCTIONSET *that are defined by Δ-formulas, then the set*

$$\text{SET} = \{\ulcorner s \urcorner \mid s \in \mathit{Set}\}$$

is recursive, and has a Δ-definition Set.

> *Chaff:* Try to keep the various typefaces straight:
>
> - *Set* is a bunch of $\mathcal{L}_{NT}$-expressions—strings of symbols from $\mathcal{L}_{NT}$.

- SET is a set of natural numbers—the Gödel numbers of the strings in *Set*.
- *Set* is an $\mathcal{L}_{NT}$-formula such that $\mathfrak{N} \models Set(\overline{a})$ if and only if there is an $s \in Set$ such that $\ulcorner s \urcorner = a$.

Proof: We follow very closely the proof of the recursiveness of the set TERM, which begins on page 152. As you worked through Exercise 6 in Section 4.8.1, you saw that you could prove the analogs of Lemmas 4.8.4 through 4.8.6 for formulas, so you have established the following lemma:

Lemma 4.10.3. *Suppose that s is an $\mathcal{L}_{NT}$-expression and u is a substring of s that is also an expression. Then*

1. *If $\ulcorner s \urcorner = a$, then the number of symbols in s is less than or equal to a.*
2. *The length of the shortest construction sequence of s is less than or equal to the number of symbols in s.*
3. $\ulcorner u \urcorner < \ulcorner s \urcorner$.

Now we can write a Δ-definition of SETCONSTRUCTIONSEQUENCE and then use this lemma to show SET is recursive:

$$
\begin{gathered}
\textit{SetConstructionSequence}(c, a) \text{ is:} \\
\textit{CodeNumber}(c) \wedge \\
(\exists l < c)\Big[\textit{Length}(c, l) \wedge \textit{IthElement}(a, l, c) \wedge \\
(\forall i \leq l)(\forall e < c)\Big(\textit{IthElement}(e, i, c) \rightarrow \\
\textit{BaseCaseSet}(e) \vee \\
(\exists j < i)(\exists e_j < c) \\
\big(\textit{IthElement}(e_j, j, c) \wedge \textit{ConstructionSet}(e_j, e)\big)\Big)\Big].
\end{gathered}
$$

We know, by assumption, that there are Δ-formulas *BaseCaseSet* and *ConstructionSet* that define the recursive sets BASECASESET and CONSTRUCTIONSET. Thus, all of the quantifiers in the definition above are bounded, and *SetConstructionSequence* is a Δ-formula.

As before we can use Lemmas 4.8.5 and 4.8.7 to bound the size of the shortest construction sequence for a: By the same argument as on page 155, there is a construction sequence coded by a number c such that $c < \left(2^{a^a}\right)^{a^2}$. So we define

$Set(a)$ is: $$\left(\exists c < \left(\overline{2}^{a^a}\right)^{a^{\overline{2}}}\right) SetConstructionSequence(c, a)$$

and we have a Δ-definition of the set SET, finishing our proof. ■

This is a wonderful theorem, as it saves us lots of work. Merely by noting that their definitions fit the requirements of Theorem 4.10.2, the following sets all turn out to be recursive and have Δ-definitions:

1. FREE, where $(x, f) \in$ FREE if and only if x is the Gödel number of a variable that is free in the formula with Gödel number f.

2. SUBSTITUTABLE, where $(t, x, f) \in$ SUBSTITUTABLE if and only if t is the Gödel number of a term that is substitutable for the variable with Gödel number x in the formula with Gödel number f.

4.10.1 Exercises

1. Work through the details (including the needed modification of Theorem 4.10.2) and show that both FREE and SUBSTITUTABLE are recursive.

2. Suppose that the function $f : \mathbb{N}^{k+1} \to \mathbb{N}$ is defined as follows:

$$f(\underset{\sim}{a}, 0) = g(\underset{\sim}{a})$$
$$f(\underset{\sim}{a}, b+1) = h(\underset{\sim}{a}, b, f(\underset{\sim}{a}, b)),$$

where g and h are recursive functions represented by Δ-formulas. Show that f is a recursive function. [*Suggestion:* One approach is to define the formula $f\,ConstructionSequence(c, \underset{\sim}{a}, l)$, with the idea that the sequence coded by c will be $\langle f(0), f(1), \ldots f(l)\rangle$. Or, you might try to fit this situation into a theorem along the lines of Theorem 4.10.2.]

3. Use Exercise 2 to show that the following functions are recursive:

   (a) The factorial function $n!$

   (b) The Fibonacci function F, where $F(1) = F(2) = 1$, and for $k \geq 3$, $F(k) = F(k-1) + F(k-2)$

   (c) The function $a \uparrow i$, where $a \uparrow 0 = 1$ and $a \uparrow (j+1) = a^{a \uparrow j}$ (You should also compute a few values, along the lines of $2 \uparrow 3$, $2 \uparrow 4$, and $2 \uparrow 5$.)

4.11 The Collection of Axioms Is Recursive

In this section we will exhibit two Δ-formulas that are designed to pick out the axioms of our deductive system.

Proposition 4.11.1. *The collection of Gödel numbers of the axioms of N is recursive.*

Proof: The formula *AxiomOfN* is easy to describe. As there are only a finite number of N-axioms, a natural number a is in the set AxiomOfN if and only if it is one of a finite number of Gödel numbers. Thus

$$
\begin{gathered}
\mathit{AxiomOfN}(a) \text{ is:} \\
a = \overline{\ulcorner (\forall x) \neg Sx = 0 \urcorner} \quad \lor \\
a = \overline{\ulcorner (\forall x)(\forall y)\left[Sx = Sy \rightarrow x = y\right] \urcorner} \quad \lor \\
\vdots \\
\lor\ a = \overline{\ulcorner (\forall x)(\forall y)\left[(x < y) \lor (x = y) \lor (y < x)\right] \urcorner}.
\end{gathered}
$$

(To be more-than-usually picky, we need to change the x's and y's to v_1's and v_2's, but you can do that.) ■

Proposition 4.11.2. *The collection of Gödel numbers of the logical axioms is recursive.*

Proof: The formula that recognizes the logical axioms is more complicated than the formula *AxiomOfN* for two reasons. The first is that there are infinitely many logical axioms, so we cannot just list them all. The second reason that this group of axioms is more complicated is that the quantifier axioms depend on the notion of substitutability, so we will have to use our results from Section 4.10.

Quite probably you are at a point where you could turn to Section 2.3.3 and write down a Δ-definition of the set LogicalAxiom. To do so would be a worthwhile exercise. But if you are feeling lazy, here is an attempt:

LogicalAxiom(a) is:

$$(\exists x < a)(\mathit{Variable}(x) \wedge a = \bar{2}^{\bar{7}}\bar{3}^{x}\bar{5}^{x}) \vee$$

$$(\exists x, y < a)\Bigg(\mathit{Variable}(x) \wedge \mathit{Variable}(y) \wedge$$

$$a = \bar{2}^{\bar{3}}\bar{3}^{(\bar{2}^{\bar{1}}\bar{3}^{(\bar{2}^{\bar{7}}\bar{3}^{x}\bar{5}^{y})})}\bar{5}^{(\bar{2}^{\bar{7}}\bar{3}^{\bar{2}^{\overline{11}}\bar{3}^{x}}\bar{5}^{\bar{2}^{\overline{11}}\bar{3}^{y}})}\Bigg) \vee$$

$$\Bigg((\exists x_1, x_2, y_1, y_2 < a)(\mathit{Variable}(x_1) \wedge \cdots \wedge \mathit{Variable}(y_2) \wedge$$

$$a = \text{Ugly Mess saying}\big[(x_1 = y_1) \wedge (x_2 = y_2)\big] \rightarrow$$

$$(x_1 + x_2 = y_1 + y_2))\Bigg) \vee$$

$$\vdots$$

(Similar clauses coding up (E2) and (E3) for $\cdot$, E, $=$, and $<$)

$$\vdots$$

$$\vee\ (\exists f, x, t, y < a)\Bigg(\mathit{Formula}(f) \wedge \mathit{Variable}(x) \wedge \mathit{Term}(t) \wedge$$

$$\mathit{Substitutable}(t, x, f) \wedge \mathit{Sub}(f, x, t, y) \wedge$$

$$a = \bar{2}^{\bar{3}}\bar{3}^{\bar{2}^{\bar{1}}\bar{3}^{(\bar{2}^{\bar{5}}\bar{3}^{x}\bar{5}^{f})}}\bar{5}^{y}\Bigg) \vee$$

(Similar clause for Axiom (Q2)).

To look at this in a little more detail, the first clause of the formula is supposed to correspond to axiom (E1): $x = x$ for each variable x. So a is the Gödel number of an axiom of this form if there is some x that is the *Gödel number of* a variable [which is what *Variable*(x) says] such that a is the *Gödel number of a formula* that looks like

variable with Gödel number x = variable with Gödel number x.

But the Gödel number for this formula is just $2^7 3^x 5^x$, so that is what we demand that a equal.

The second clause covers axioms of the form (E2), when the function f is the function S. We demand that a be the code for the formula

$$(v_i = v_j) \rightarrow Sv_i = Sv_j,$$

where $x = \ulcorner v_i \urcorner$ and $y = \ulcorner v_j \urcorner$. After you fuss with the coding, you come out with the expression shown. The other clauses of type (E2) and (E3) are similar.

The last clause that is written out is for the quantifier axiom (Q1). After demanding that the term coded by t be substitutable for the variable coded by x in the formula coded by f and that y be the code for the result of substituting in that way, the equation for a is nothing more than the analog of $(\forall x \phi) \rightarrow \phi_t^x$. ■

4.11.1 Exercise

1. Complete the definition of the formula *LogicalAxiom*.

4.12 Coding Deductions

It is probably difficult to remember at this point of our journey, but our goal is to prove the Incompleteness Theorem, and to do that we need to write down an $\mathcal{L}_{NT}$-sentence that is true in $\mathfrak{N}$, the standard structure, but not provable from the axioms of N. Our sentence, θ, will "say" that θ is not provable from N, and in order to "say" that, we will need a formula that will identify the (Gödel numbers of the) formulas that *are* provable from N. To do that we will need to be able to code up deductions from N, which makes it necessary to code up sequences of formulas. Thus, our next goal will be to settle on a coding scheme for sequences of $\mathcal{L}_{NT}$-formulas.

We have been pretty careful with our coding up to this point. If you check, every Gödel number that we have used has been even, with the exception of 3, which is the garbage case in Definition 4.7.1. We will now use numbers with smallest prime factor 5 to code sequences of formulas.

Suppose that we have the sequence of formulas

$$D = \langle \phi_1, \phi_2, \ldots, \phi_k \rangle.$$

We will define the **sequence code of** D to be the number

$$\ulcorner D \urcorner = 5^{\ulcorner \phi_1 \urcorner} 7^{\ulcorner \phi_2 \urcorner} \cdots p_{k+2}^{\ulcorner \phi_k \urcorner}.$$

So the exponent on the $(i+2)$nd prime is the Gödel number of the ith element of the sequence. You are asked in the Exercises to produce several useful $\mathcal{L}_{NT}$-formulas relating to sequence codes.

We will be interested in using sequence codes to code up deductions from N. If you look back at the definition of a deduction (Definition 2.2.1), you will see that to check if a sequence is a deduction, we need only check that each entry is either an axiom or follows from previous lines of the deduction via a rule of inference. So to say that c codes up a deduction from N, we want to be able to say, for each entry e at position i of the deduction coded by c,

$$AxiomOfN(e) \lor LogicalAxiom(e) \lor RuleOfInference(c, e, i).$$

The first two of these we have already developed. The last major goal of this section (and this chapter) is to flesh out the details of a Δ-definition of a formula that recognizes when an entry in a deduction is justified by one of our rules of inference.

As you recall from Section 2.4, there are two types of rules of inference: propositional consequence and quantifier rules. The latter of these is easiest to Δ-define, so we deal with it first.

The rule of inference (QR) is Definition 2.4.6, which says that if x is not free in ψ, then the following are rules of inference:

$$\langle \{\psi \rightarrow \phi\}, \psi \rightarrow (\forall x \phi) \rangle$$
$$\langle \{\phi \rightarrow \psi\}, (\exists x \phi) \rightarrow \psi \rangle.$$

If we look at the first of these, we see that if e is an entry in a code for a deduction, and $e = \ulcorner \psi \rightarrow (\forall x \phi) \urcorner$, then e is justified as long as there is an earlier entry in the deduction that codes up the formula $\psi \rightarrow \phi$, assuming that x is not free in ψ. So all we have to do is figure out a way to say this.

We will write the formula *QRRule1*(c, e, i), where c is the code of the deduction, and e is the entry at position i that is being justified by the first quantifier rule. In the following, f is playing the role of $\ulcorner \psi \urcorner$, while g is supposed to be $\ulcorner \phi \urcorner$:

$$
\begin{aligned}
&\mathit{QRRule1}(c, e, i) \text{ is:} \\
&\qquad \mathit{SequenceCode}(c) \land \mathit{IthSequenceElement}(e, i, c) \land \\
&(\exists x, f, g < c)\Big[\mathit{Formula}(f) \land \mathit{Formula}(g) \land \mathit{Variable}(x) \land \\
&\qquad \neg \mathit{Free}(x, f) \land \\
&\qquad e = \overline{2}^{\overline{3}} \overline{3}^{\overline{2}^{\overline{1}} \overline{3}^{f}} \overline{5}^{\overline{2}^{\overline{5}} \overline{3}^{x} \overline{5}^{g}} \land \\
&\qquad (\exists j < i)(\exists e_j < c)\big(\mathit{IthSequenceElement}(e_j, j, c) \land \\
&\qquad\qquad e_j = \overline{2}^{\overline{3}} \overline{3}^{\overline{2}^{\overline{1}} \overline{3}^{f}} \overline{5}^{g}\big)\Big].
\end{aligned}
$$

After you write out **QRRule2** in the Exercises, it is obvious to define

$$
\begin{aligned}
&\mathit{QRRule}(c, e, i) \text{ is:} \\
&\qquad \mathit{QRRule1}(c, e, i) \lor \mathit{QRRule2}(c, e, i).
\end{aligned}
$$

Now we have to address propositional consequence. This will involve some rather tricky coding, so hold on tight as we review propositional logic.

Assume that D is our deduction, and D is the sequence of formulas

$$\langle \alpha_1, \alpha_2, \alpha_3, \ldots, \alpha_k \rangle.$$

Notice that entry α_i of a deduction is justified as a propositional consequence if and only if the formula

$$\beta = (\alpha_1 \land \alpha_2 \land \cdots \land \alpha_{i-1}) \rightarrow \alpha_i$$

is a tautology.

Now, as we discussed in Section 2.4, in order to decide if a first-order formula β is a tautology, we must take β and find the propositional formula β_P. Then if β_P is

$$\Big[(\alpha_1)_P \land (\alpha_2)_P \land \cdots \land (\alpha_{i-1})_P\Big] \rightarrow (\alpha_i)_P,$$

we must show any truth assignment that makes $(\alpha_1)_P$ through $(\alpha_{i-1})_P$ true must also make $(\alpha_i)_P$ true.

As outlined on page 59, to create a propositional version of a formula, we first must find the prime components of that formula, where a prime component is a subformula that is either universal and not contained in any other universal subformula, or atomic and not contained in any universal formula. Rather than explicitly writing out a Δ-formula that identifies the pairs (u, v) such that u is the Gödel number of a prime component of the formula with Gödel number v, let us write down a recursive definition of this set of formulas, and we will leave it to the reader to find the minor modification of Theorem 4.10.2 which will guarantee that the set of Gödel numbers is recursive.

Definition 4.12.1. If β and γ are $\mathcal{L}_{NT}$-formulas, γ is said to be a **prime component** of β if:

1. β is atomic and $\gamma = \beta$, or
2. β is universal and $\gamma = \beta$, or
3. β is $\neg\alpha$ and γ is a prime component of α, or
4. β is $\alpha_1 \lor \alpha_2$ and γ is a prime component of either α_1 or α_2.

Proposition 4.12.2. *The set*

$$\begin{aligned}&\text{PRIMECOMPONENT} = \\ &\{(u, f) \mid u = \ulcorner\gamma\urcorner \textit{ and } f = \ulcorner\beta\urcorner \textit{ and } \gamma \textit{ is a prime component of } \beta,\\ &\qquad\qquad\qquad\qquad\textit{for some } \mathcal{L}_{NT}\textit{-formulas } \gamma \textit{ and } \beta\}\end{aligned}$$

is recursive and has Δ-definition $\boxed{PrimeComponent(u, f)}$.

Proof: Theorem 4.10.2. ■

Now we will code up a canonical sequence of all of the prime components of α_1 through α_i. We will say r codes the PrimeList for the first i entries of the deduction coded by c if each element coded by r is a prime component of one of the first i entries of the deduction coded by c, r's elements are distinct, each prime component of each of the first i entries of the deduction coded by c is among the entries in r, and if s is a smaller code number, s is missing one of these prime components:

$$
\begin{array}{l}
PrimeList(c,i,r) \text{ is:}\\
\qquad SequenceCode(c) \land CodeNumber(r) \land\\
\qquad (\exists l < r)\Big[Length(r,l) \land\\
\qquad (\forall m, n \leq l)(\forall e_m, e_n < r)\\
\qquad (IthElement(e_m, m, r) \land IthElement(e_n, n, r)) \rightarrow\\
\qquad \Big((\exists k \leq i)(\exists f_k \leq c) IthSequenceElement(f_k, k, c) \land\\
\qquad PrimeComponent(e_m, f_k) \land\\
\qquad [(m \neq n) \rightarrow (e_m \neq e_n)]\Big) \land\\
\qquad [(\forall k \leq i)(\forall f_k \leq c)(\forall u \leq f_k)(IthSequenceElement(f_k, k, c) \land\\
\qquad PrimeComponent(u, f_k) \rightarrow\\
\qquad (\exists m < l) IthElement(u, m, r))]\Big] \land\\
\qquad (\forall s < r)\Big(CodeNumber(s) \rightarrow\\
\qquad (\exists k \leq i)(\exists f_k \leq c)(\exists u \leq f_k)\big[IthSequenceElement(f_k, k, c) \land\\
\qquad PrimeComponent(u, f_k) \land\\
\qquad (\forall m < s)(\neg IthElement(u, m, s))\big]\Big)
\end{array}
$$

To decide if β is a tautology, we need to assign all possible truth values to all of the prime components in our list, and then evaluate the truth of each α_i under a given truth assignment. So the next thing we need to do is find a way to code up an assignment of truth values to all of the prime components of all the α_i's. We will say that v codes up a truth assignment if v is the code number of a sequence of the right length and all of the elements coded by v are either 1 (for false) or 2 (for true). [Although it would be convenient to use 0's and 1's in v, that runs afoul of our convention that we never code a 0.]

$$\begin{aligned}
&\mathit{TruthAssignment}(c,i,r,v) \text{ is:}\\
&\qquad \mathit{PrimeList}(c,i,r) \land \mathit{CodeNumber}(v) \land\\
&\qquad (\exists l < r)\big(\mathit{Length}(r,l) \land \mathit{Length}(v,l) \land\\
&\qquad (\forall i \leq l)(\forall e < v)(\mathit{IthElement}(e,i,v) \rightarrow\\
&\qquad\qquad [e = \overline{1} \lor e = \overline{2}])\big).
\end{aligned}$$

Now, given a truth assignment for the prime components of α_1 through α_i, coded up in v, we need to be able to evaluate the truth of each formula α_n under that assignment. To do this we will need to be able to evaluate the truth value of a single formula.

Suppose that we first look at an example. Here is a formula construction sequence that ends with some formula α:

$$\big\langle 0 < x, x < y, (0 < x \lor x < y), \neg(0 < x),\\ (\forall x)(0 < x \lor x < y), \underbrace{(\forall x)(0 < x \lor x < y) \lor (\neg 0 < x)}_{\alpha}\big\rangle.$$

Let us assume that the PrimeList we are working with is

$$\langle (\forall x)(0 < x \lor x < y), 0 < x \rangle$$

and the chosen truth assignment for our PrimeList is

$$\langle 1, 2 \rangle.$$

To assign the truth value to α, we follow along the construction sequence. When we see an entry that is in the PrimeList, we assign that entry the corresponding truth value from our truth assignment. If the entry in the construction sequence is not a prime component, one of three things might be true:

1. The entry might be universal, in which case we assign it truth value 3 (for *undefined*).

2. The entry might be an atomic formula that ends up inside the scope of a quantifier in α. Again, we use truth value 3.

3. The entry might be the denial of or disjunction of earlier entries in the construction sequence. In this case we can figure out its truth value, always using 3 if any of the parts have truth value 3.

So to continue our example from above, the sequence of truth values would be

$$\langle 2, 3, 3, 1, 1, 1 \rangle.$$

Exercise 7 asks you to write a Δ-formula $\boxed{Evaluate(e, r, v, y)}$, where you should assume that e is the Gödel number for a formula α, r is a code for a list including all of the prime components of α, v is a TruthAssignment for r, and y is the truth value for α, given the truth assignment v. Exercise 8 also concerns this formula.

Now, knowing how to evaluate the truth of a single formula α, we will be able to decide if α_i, the ith element of the alleged deduction that is coded by c, can be justified by the propositional consequence rule. Recall that we need only check whether

$$(\alpha_1 \wedge \alpha_2 \wedge \cdots \wedge \alpha_{i-1}) \rightarrow \alpha_i$$

is a tautology. To do this, we need only see whether any truth assignment that makes α_1 through α_{i-1} true also makes α_i true. Here is the Δ-formula that says this, where c codes the alleged deduction, and e is supposed to be the code for the ith entry in the deduction, the entry that is to be justified by an appeal to the rule PC:

$PCRule(c, e, i)$ is:

$$IthSequenceElement(e, i, c) \wedge$$

$$\left(\exists r < \left[\overline{2}^{\left(c^{\overline{2}}\right)^{\left(c^{\overline{2}}\right)}}\right]^{c^{\overline{3}}}\right)$$

$$\left(\forall v < \left(\left[\overline{2}^{\left(c^{\overline{2}}\right)^{\left(c^{\overline{2}}\right)}}\right]^{\ulcorner \overline{2} \urcorner}\right)^{c^{\overline{2}}}\right)$$

$$\Big(\left[PrimeList(c, i, r) \wedge TruthAssignment(c, i, r, v)\right] \rightarrow$$

$$\Big[\big((\forall j < i)(\exists e_j < c)$$

$$(IthSequenceElement(e_j, j, c) \wedge Evaluate(e_j, r, v, \overline{2}))\big)$$

$$\rightarrow$$

$$Evaluate(e, r, v, \overline{2})\Big]\Big).$$

Now, to know that this works, we must justify the bounds that we have given for r and v. The number r is supposed to code the list of prime components among the first i elements of the deduction c. So r is of the form $5^{\ulcorner \gamma_1 \urcorner} 7^{\ulcorner \gamma_2 \urcorner} \cdots p_{k+2}^{\ulcorner \gamma_k \urcorner}$, where the prime components are γ_1 through γ_k. First, we need to get a handle on the number of prime components there are. Since c is the code for the deduction, there are fewer than c formulas in the deduction, and each of those formulas has a Gödel number that is less than or equal to c. So each formula in the deduction has fewer than c symbols in it, and thus fewer than c prime components. So we have no more than c formulas, each with no more than c prime components, so there are no more than c^2 prime components total in the deduction coded by c. If we look at the number r, we see that

$$\begin{aligned} r &= 5^{\ulcorner \gamma_1 \urcorner} 7^{\ulcorner \gamma_2 \urcorner} \cdots p_{k+2}^{\ulcorner \gamma_k \urcorner} \\ &\leq 5^c 7^c \cdots p_{k+2}^c \\ &\leq 5^c 7^c \cdots p_{c^2}^c \\ &\leq \left[2^{(c^2)^{(c^2)}} \right]^{c^3}, \end{aligned}$$

where the last line depends on our usual bound for the size of the (c^2)th prime, as we saw on page 155.

As for v, v is a code number that gives us the truth values of the various prime components coded in r. Thus v is of the form $2^{i_1} 3^{i_2} \cdots p_k^{i_k}$, where there are k prime components, and each i_j is either $\ulcorner \bar{1} \urcorner$ or $\ulcorner \bar{2} \urcorner$. By the argument above, we know there are no more than c^2 prime components, so

$$\begin{aligned} v &= 2^{i_1} 3^{i_2} \cdots p_k^{i_k} \\ &\leq 2^{\ulcorner \bar{2} \urcorner} 3^{\ulcorner \bar{2} \urcorner} \cdots p_{c^2}^{\ulcorner \bar{2} \urcorner} \\ &\leq \left(\left[2^{(c^2)^{(c^2)}} \right]^{\ulcorner \bar{2} \urcorner} \right)^{c^2}. \end{aligned}$$

Thus we have justified the bounds given for r and v in the definition of $PCRule(c, e, i)$. Thus we have a Δ-definition, and the set PCRule is recursive.

Now we have to remember where we were on page 171. We are thinking of c as coding an alleged deduction from N, and we need to check all of the entries of c to see if they are legal. We have already written Δ-formulas *LogicalAxiom* and *AxiomOfN*, and we are working on the rules of inference. The quantifier rules were relatively easy, so we then wrote a Δ-formula $PCRule(c, e, i)$ that is true if and only if e is the ith entry of the deduction c and can be justified by the rule PC.

At last, we are able to decide if c is the code for a deduction from N of a formula with Gödel number f:

$$\begin{aligned}
&\mathit{Deduction}(c, f) \text{ is:}\\
&\qquad \mathit{SequenceCode}(c) \wedge \mathit{Formula}(f) \wedge\\
&(\exists l < c)\Big(\mathit{SequenceLength}(c, l) \wedge \mathit{IthSequenceElement}(f, l, c) \wedge\\
&\qquad (\forall i \leq l)(\forall e < c)\Big[\mathit{IthSequenceElement}(e, i, c) \rightarrow\\
&\qquad\qquad \Big(\mathit{LogicalAxiom}(e) \vee \mathit{AxiomOfN}(e) \vee\\
&\qquad\qquad\qquad \mathit{QRRule}(c, e, i) \vee \mathit{PCRule}(c, e, i)\Big)\Big]\Big).
\end{aligned}$$

This formula represents the set DEDUCTION $\subseteq \mathbb{N}^2$ and shows that DEDUCTION is a recursive set. This makes formal the ideas of Chapter 2, where it was suggested that we ought to be able to program a computer to decide if an alleged deduction is, in fact, a deduction. If you keep the equivalence between recursive and "computer-decidable" in your head, we will say more about this in Chapter 5.

4.12.1 Exercises

1. Write a Δ-formula $\boxed{\mathit{SequenceCode}(c)}$ that is true if and only if c is the code of a sequence of $\mathcal{L}_{NT}$-formulas.

2. Write a Δ-formula $\boxed{\mathit{SequenceLength}(c, l)}$ that is true if and only if c is a sequence code of a sequence of length l.

3. Write out a Δ-formula $\boxed{\mathit{IthSequenceElement}(e, i, c)}$ that is true in $\mathfrak{N}$ if and only if c is a sequence code and the ith element of the sequence coded by c has Gödel number e.

4. Write out a Δ-formula $QRRule2(c, e, i)$ that will be true if e is the ith entry in the sequence coded by c and is justified by the second quantifier rule.

5. Here is your average, ordinary tautology:

$$\phi(x, y) \text{ is } \big[[\forall x P(x)] \rightarrow (Q(x, y) \rightarrow [\forall x P(x)])\big].$$

Find a construction sequence for ϕ. Make a list of the prime components of ϕ. Pretending that your list of prime components is *the* prime list for ϕ, find all possible truth assignments for ϕ and use the truth assignments to evaluate the truth of ϕ under your assignments. If all goes well, every time you evaluate the truth of ϕ, you will get: True.

6. Repeat Exercise 5 with the following formulas, which are not (necessarily) guaranteed to be tautologies:

   (a) $(\forall x)(x < y \rightarrow x < y)$
   (b) $(\forall x)(x < y) \rightarrow (\forall x)(x < y)$
   (c) $(A(x) \vee B(y)) \vee \neg B(x)$
   (d) $(A(x) \vee B(y)) \rightarrow \big[(\forall x)(\forall y)(A(x) \vee B(y))\big]$
   (e) $\big[(\forall x)(\forall y)(A(x) \vee B(y))\big] \rightarrow (A(x) \vee B(y))$

7. Write out the Δ-formula $Evaluate(e, r, v, y)$, as outlined in the text. You will need to think about formula construction sequences, as you did in Exercise 6 in Section 4.8.1.

8. Show by induction on the longest of the two construction sequences that if d_1 and d_2 are codes for two construction sequences of α, and if $Evaluate(d_1, r, v, y_1)$ and $Evaluate(d_2, r, v, y_2)$ are both true, then y_1 and y_2 are equal. Thus, the truth assigned to α does not depend upon the construction sequence chosen to evaluate that truth.

4.13 Summing Up, Looking Ahead

Well, in all likelihood you are exhausted at this point. This chapter has been full of dense, technical arguments with imposing definition piled upon imposing definition. We have established our axioms, discussed recursive sets, and talked about Δ-definitions. You have just finished wading through an unending stream of Δ-definitions that culminated with the formula $Deduction(c, f)$ which holds if and only if c is a code for a deduction of the formula with Gödel number f. We have succeeded in coding up our deductive theory inside of number theory.

Let me reiterate this. If you look at that formula *Deduction*, what it looks like is a disjunction of a lot of equations and inequalities. Everything is written in the language $\mathcal{L}_{NT}$, so *everything* in that formula is of the form $SSS0 < SS0 + x$ (with, it must be admitted, rather more S's than shown here). Although we have given these formulas names which suggest that they are about formulas and terms and tautologies and deductions, the formulas are formulas of elementary number theory, so the formulas don't know that they are about anything beyond whether this number is bigger than that number, no matter how much you want to anthropomorphize them. The interpretation of the numbers as standing for formulas via the scheme of Gödel numbering is imposed on those numbers by us.

The next chapter brings us to the statement and the proof of Gödel's Incompleteness Theorem. To give you a taste of things to come, notice that if we define the statement

$Thm_N(f)$ is:

$$(\exists c)(Deduction(c, f)),$$

then $Thm_N(f)$ should hold if and only if f is the Gödel number of a formula that is a theorem of N. I am sure you noticed that Thm_N is not a Δ-formula, and there is no way to fix that—we cannot bound the length of a deduction of a formula. But Thm_N *is* a Σ-formula, and Proposition 4.3.9 tells us that true Σ-sentences are provable. That will be one of the keys to Gödel's proof.

Well, if 90% of the iceberg is under water, we've covered that. Now it is time to examine that glorious 10% that is left.

Chapter 5

The Incompleteness Theorems

5.1 Introduction

Suppose that A is a collection of axioms in the language of number theory such that A is consistent and is simple enough so that we can decide whether or not a given formula is an element of A. The First Incompleteness Theorem will produce a sentence, θ, such that $\mathfrak{N} \models \theta$ and $A \nvdash \theta$, thus showing our collection of axioms A is incomplete.

The idea behind the construction of θ is really neat: We get θ to say that θ is not provable from the axioms of A. In some sense, θ is no more than a fancy version of the liar's paradox, in which the speaker asserts that the speaker is lying, inviting the listener to decide whether that utterance is a truth or a falsehood. The challenge for us is to figure out how to get an $\mathcal{L}_{NT}$-sentence to do the asserting!

You will notice that there are two parts to θ. The first is that θ will have to talk about the collection of Gödel numbers of theorems of A. That is no problem, as we will have a Σ-formula $Thm_A(f)$ that is true (and thus provable from N) if and only if f is the Gödel number of a theorem of A. The thing that makes θ tricky is that we want θ to be $Thm_A(\overline{a})$, where $a = \ulcorner \theta \urcorner$. In this sense, we need θ to refer to itself. Showing that we can do that will be the content of the Self-Reference Lemma that we address in the next section.

After proving the First Incompleteness Theorem, we will discuss some corollaries to the theorem before moving on to discuss the Second Incompleteness Theorem, which states that the set of axioms of Peano Arithmetic cannot prove that Peano Arithmetic is consistent, unless (of course) Peano Arithmetic is inconsistent, in which case it can prove anything. So our goal of proving that we have a complete, consistent set of axioms for $\mathfrak{N}$ is a goal that cannot be reached within the confines of first-order logic.

5.2 The Self-Reference Lemma

Given any formula with only one free variable, we show how to construct a sentence that asserts that the given formula applies to itself:

Lemma 5.2.1 (Gödel's Self-Reference Lemma). *Let $\psi(v_1)$ be an $\mathcal{L}_{NT}$-formula with only v_1 free. Then there is a sentence ϕ such that*

$$N \vdash \left(\phi \leftrightarrow \psi(\ulcorner \phi \urcorner)\right).$$

> *Chaff:* Look at how neat this is! Do you see how ϕ "says" ψ *is true of me*? And we can do this for any formula ψ! What a cool idea!

Proof: We will explicitly construct the needed ϕ. Recall that we have recursive functions Num : $\mathbb{N} \to \mathbb{N}$ and Sub : $\mathbb{N}^3 \to \mathbb{N}$ such that $\text{Num}(n) = \ulcorner \overline{n} \urcorner$ and $\text{Sub}(\ulcorner \alpha \urcorner, \ulcorner x \urcorner, \ulcorner t \urcorner) = \ulcorner \alpha^x_t \urcorner$. Since these functions are represented by our Δ-definitions of Section 4.9, we know that

$$N \vdash \left[Num(\overline{a}, y) \leftrightarrow y = \overline{\text{Num}(a)}\right], \text{ and that}$$
$$N \vdash \left[Sub(\overline{a}, \overline{b}, \overline{c}, z) \leftrightarrow z = \overline{\text{Sub}(a, b, c)}\right].$$

> *Chaff:* Remember, Num is the function and *Num* is the $\mathcal{L}_{NT}$-formula that represents the function! Oh, and just because we're going to need it, $\ulcorner v_1 \urcorner = 4$.

Now suppose that $\psi(v_1)$ is given as in the statement of the lemma. Let $\gamma(v_1)$ be

$$\forall y \forall z \left[\left[Num(v_1, y) \wedge Sub(v_1, \overline{4}, y, z)\right] \rightarrow \psi(z)\right].$$

Let us look at $\gamma(n)$ a little more closely, supposing that $n = \ulcorner \alpha \urcorner$. If the antecedent of $\gamma(n)$ holds, then the first part of the antecedent tells us that

$$y = \text{Num}(n) = \ulcorner \overline{n} \urcorner$$

and the second part of the antecedent asserts that

$$\begin{aligned} z &= \mathrm{Sub}(\overline{n}, \overline{4}, \ulcorner \overline{n} \urcorner) \\ &= \mathrm{Sub}(\overline{\ulcorner \alpha \urcorner}, \overline{\ulcorner v_1 \urcorner}, \ulcorner \overline{n} \urcorner) \\ &= \ulcorner \alpha^{v_1}_{\overline{n}} \urcorner \\ &= \ulcorner \alpha^{v_1}_{\overline{\ulcorner \alpha \urcorner}} \urcorner. \end{aligned}$$

So z is the Gödel number of $\{\alpha$ with the Gödel number of α substituted in for $v_1\}$.

One more tricky choice will get us to where we want to go. Let $m = \ulcorner \gamma(v_1) \urcorner$, and let ϕ be $\gamma(\overline{m})$. Certainly, ϕ is a sentence, so we will be finished if we can show that $N \vdash \phi \leftrightarrow \psi(\overline{\ulcorner \phi \urcorner})$.

Let us work through a small calculation first. Notice that

$$\begin{aligned} \mathrm{Sub}(m, 4, \ulcorner \overline{m} \urcorner) &= \mathrm{Sub}(\ulcorner \gamma(v_1) \urcorner, \ulcorner v_1 \urcorner, \ulcorner \overline{m} \urcorner) \\ &= \ulcorner \gamma(v_1)^{v_1}_{\overline{m}} \urcorner \\ &= \ulcorner \gamma(\overline{m}) \urcorner \\ &= \ulcorner \phi \urcorner. \end{aligned} \tag{5.1}$$

With this in hand, the following are provably equivalent in N:

ϕ

$\forall y \forall z \Big[Num(\overline{m}, y) \rightarrow \big(Sub(\overline{m}, \overline{4}, y, z) \rightarrow \psi(z) \big) \Big]$ — logic

$\forall y \forall z \Big[y = \overline{\mathrm{Num}(m)} \rightarrow \big(Sub(\overline{m}, \overline{4}, y, z) \rightarrow \psi(z) \big) \Big]$ — *Num* represents Num

$\forall y \forall z \Big[y = \overline{\ulcorner \overline{m} \urcorner} \rightarrow \big(Sub(\overline{m}, \overline{4}, y, z) \rightarrow \psi(z) \big) \Big]$ — calculation

$\forall z \Big(Sub(\overline{m}, \overline{4}, \overline{\ulcorner \overline{m} \urcorner}, z) \rightarrow \psi(z) \Big)$ — quantifier rules

$\forall z \Big(z = \overline{\mathrm{Sub}(m, 4, \ulcorner \overline{m} \urcorner)} \rightarrow \psi(z) \Big)$ — *Sub* represents Sub

$\forall z \Big(z = \overline{\ulcorner \phi \urcorner} \rightarrow \psi(z) \Big)$ — calculation (5.1) above

$\psi(\overline{\ulcorner \phi \urcorner})$ — quantifier rules

So $N \vdash \phi \leftrightarrow \psi\Big(\overline{\ulcorner \phi \urcorner}\Big)$, as needed. ■

Notice in this proof that if ψ is a Π-formula, then ϕ is logically equivalent to a Π-sentence. By altering γ slightly we can also arrange, if ψ is a Σ-formula, to have ϕ logically equivalent to a Σ-sentence.

5.2.1 Exercises

1. Suppose that $\psi(v_1)$ is $Formula(v_1)$. By the Self-Reference Lemma, there is a sentence ϕ such that $N \vdash (\phi \leftrightarrow Formula(\ulcorner\phi\urcorner))$. Does $N \vdash \phi$? Does $N \vdash \neg\phi$? Justify your answer. What happens if we use $\psi(v_1) = \neg Formula(v_1)$ instead?

2. Let $\psi(v_1)$ be $Even(v_1)$, and let ϕ be the sentence generated when the Self-Reference Lemma is applied to $\psi(v_1)$. Does $N \vdash \phi$? Does $N \vdash \neg\phi$? How can you tell?

3. Show that the proof of the Self-Reference Lemma still works if we use

$$\gamma(v_1) = \exists y \exists z \Big[Num(v_1, y) \land Sub(v_1, \overline{4}, y, z) \land \psi(z) \Big].$$

Conclude that if ψ is a Σ-formula, then the ϕ of the Self-Reference Lemma can be taken to be equivalent to a Σ-sentence.

5.3 The First Incompleteness Theorem

We are ready to state and prove the First Incompleteness Theorem, which tells us that if we are given any reasonable axiom system A, there is a sentence that is true in $\mathfrak{N}$ but not provable from A.

You may have been complaining all along about my choice for an axiom system. Perhaps you have been convinced from the beginning that N is clearly too weak to prove every truth about the natural numbers. You are right. We know, for example, that N does not prove the commutative law of addition. Since you are a diligent person, I imagine that you have come up with an axiom system of your own, let's call it A. You might be convinced that A is the "right" choice of axioms, a set of axioms that is strong enough to prove every truth about $\mathfrak{N}$. Although this shows admirable independence on your part, we will, unfortunately, be able to prove that A is no better than N, as long as A satisfies certain reasonable conditions.

Definition 5.3.1. A **theory** is a collection of sentences T that is closed under deduction: For every sentence σ, if $T \vdash \sigma$, then $\sigma \in T$. If A is a set of sentences, then the **theory of** A, written $Th(A)$, is the smallest theory that includes A: $Th(A) = \{\sigma \mid A \vdash \sigma\}$.

Definition 5.3.2. A theory T in the language $\mathcal{L}_{NT}$ is said to be **recursively axiomatized** if there is a set of axioms A such that

1. $T = \{\sigma \mid \sigma \text{ is a sentence and } A \vdash \sigma\}$.

2. $\text{AXIOMOFA} \underset{\text{def}}{=} \{\ulcorner \alpha \urcorner \mid \alpha \in A\}$ is a recursive set.

Roughly, a theory is recursively axiomatized if the theory has an axiom set that is simple enough so that we can recognize axioms and nonaxioms. One of the conditions that we will require of your set of axioms A is that it be recursive. The only other condition is that A needs to be consistent.

Here is the theorem:

Theorem 5.3.3 (Gödel's First Incompleteness Theorem). *Suppose that A is a consistent and recursive set of axioms in the language $\mathcal{L}_{NT}$. Then there is a Π-sentence θ such that $\mathfrak{N} \models \theta$ but $A \nvdash \theta$.*

Proof: Although our analysis in Chapter 4 was worked out with the specific set of axioms N, it is easy to see that as long as the set AXIOMOFA is recursive, there is a recursive set DEDUCTION_A, represented by the formula $Deduction_A(c, f)$ such that c codes up an A-deduction of the formula with Gödel number f if and only if $N \vdash Deduction_A(\overline{c}, \overline{f})$.

If we let the formula $Thm_A(v_1)$ be $(\exists c)(Deduction_A(c, v_1))$, then Thm_A is a Σ-formula. Therefore, by Proposition 4.3.9, we know that if $\mathfrak{N} \models Thm_A(f)$, then for some code number c, $\mathfrak{N} \models Deduction_A(c, f)$. Therefore, $N \vdash Deduction_A(\overline{c}, \overline{f})$, as $Deduction_A(\overline{c}, \overline{f})$ is a true Δ-sentence. But this means that $N \vdash Thm_A(\overline{f})$. To summarize:

$$\text{If } \mathfrak{N} \models Thm_A(f), \text{ then } N \vdash Thm_A(\overline{f}). \tag{5.2}$$

We assume that A is strong enough to prove all of the axioms of N. If not, the universal closure of any unprovable-from-A axiom of N will be a Π-sentence that is true in $\mathfrak{N}$ and unprovable from A.

With this assumption, use the Self-Reference Lemma to produce a sentence θ such that

$$N \vdash \left[\theta \leftrightarrow \neg Thm_A\left(\overline{\ulcorner \theta \urcorner}\right)\right].$$

Chaff: Do you see how θ "says" *I am not a theorem?*

Notice by the comments following the proof of the Self-Reference Lemma that θ can be taken to be logically equivalent to a Π-sentence.

Now $\mathfrak{N} \models \left[\theta \leftrightarrow \neg Thm_A\left(\overline{\ulcorner\theta\urcorner}\right)\right]$, so we know that

$$\begin{aligned}\mathfrak{N} \models \theta &\leftrightarrow \mathfrak{N} \not\models Thm_A\left(\overline{\ulcorner\theta\urcorner}\right)\\ &\leftrightarrow \ulcorner\theta\urcorner \notin \text{THM}_A\\ &\leftrightarrow A \nvdash \theta,\end{aligned}$$

so θ is either true in $\mathfrak{N}$ and not provable from A, or false in $\mathfrak{N}$ and provable from A.

Assume, for the moment, that θ is false and $A \vdash \theta$. Then $A \vdash \theta$, so $\mathfrak{N} \models Thm_A(\ulcorner\theta\urcorner)$. By (5.2) this means that $N \vdash Thm_A\left(\overline{\ulcorner\theta\urcorner}\right)$. But then, by our choice of θ, $N \vdash \neg\theta$. Since A proves all of the axioms of N, this implies that $A \vdash \neg\theta$. But we already have assumed that $A \vdash \theta$, which means that A is inconsistent, contrary to our assumption on A.

Thus, $A \nvdash \theta$ and $\mathfrak{N} \models \theta$, and θ is true and unprovable, as needed. ∎

Corollary 5.3.4. *If A is a consistent, recursive set of axioms in the language $\mathcal{L}_{NT}$, then*

$\text{THM}_A = \{a \mid a$ *is the Gödel number of a formula derivable from* $A\}$

is not recursive.

Proof: Suppose that THM_A is recursive. Then some formula $\gamma(v_1)$ represents the set in N, which means that

$$\begin{aligned}&\text{If } f \in \text{THM}_A, \text{ then } N \vdash \gamma(\overline{f}).\\ &\text{If } f \notin \text{THM}_A, \text{ then } N \vdash \neg\gamma(\overline{f}).\end{aligned}$$

If we construct θ as in the proof of the First Incompleteness Theorem, we know that θ is either true and unprovable or false and provable. We also know that the assumption that θ is false and provable leads to a contradiction. Therefore, θ is true and unprovable. But then, since $\gamma(v_1)$ represents THM_A and $\ulcorner\theta\urcorner \notin \text{THM}_A$, $N \vdash \neg\gamma\left(\overline{\ulcorner\theta\urcorner}\right)$. But then, by our choice of θ, $N \vdash \theta$, so $A \vdash \theta$, contradicting the fact that θ is unprovable.

So our assumption that THM_A is recursive leads to a contradiction, and we are led to conclude that THM_A is not recursive. ∎

Chaff: This corollary is the "computers will never put mathematicians out of a job" corollary: If you accept the identification between recursive sets and sets for which a computer can decide membership, Corollary 5.3.4 says that we will never be able to write a computer program which will accept as input an $\mathcal{L}_{NT}$-formula ϕ and will produce as output "ϕ is a theorem" if $A \vdash \phi$ and "ϕ is not a theorem" if $A \nvdash \phi$.

If you think of the computer as taking ϕ and systematically listing all deductions and checking to see if it has listed a deduction-of-ϕ, it is easy to see that if ϕ is, in fact, a theorem-of-A, the computer will eventually verify that fact. If, however, ϕ is not a theorem, the computer will never know. All the computer will know is that it has not succeeded, as of this moment, of finding a deduction of ϕ. But it will not be able to say that it will never come across a deduction of ϕ. This is saying that the collection of theorems-of-A is recursively enumerable, but not recursive.

We can, in fact, dispense with the requirement that A be a recursive set of axioms:

Theorem 5.3.5. *Suppose that A is a consistent set of axioms extending N and in the language $\mathcal{L}_{NT}$. Then the set* THM_A *is not representable in A (and therefore* THM_A *is not recursive).*

Proof: Suppose, to the contrary, that $\gamma(v_1)$ represents THM_A. As usual, let θ be such that

$$N \vdash \left[\theta \leftrightarrow \neg\gamma\left(\overline{\ulcorner\theta\urcorner}\right)\right].$$

As A extends N, certainly

$$A \vdash \left[\theta \leftrightarrow \neg\gamma\left(\overline{\ulcorner\theta\urcorner}\right)\right].$$

Now

$$
\begin{aligned}
A \vdash \theta &\rightarrow \ulcorner\theta\urcorner \in \text{THM}_A && \\
&\rightarrow A \vdash \gamma\left(\overline{\ulcorner\theta\urcorner}\right) && \text{since } \gamma \text{ represents THM}_A \\
&\rightarrow A \vdash \neg\theta && \text{choice of } \theta \\
&\rightarrow A \nvdash \theta && A \text{ is consistent} \\
&\rightarrow \ulcorner\theta\urcorner \notin \text{THM}_A && \\
&\rightarrow A \vdash \neg\gamma\left(\overline{\ulcorner\theta\urcorner}\right) && \text{since } \gamma \text{ represents THM}_A \\
&\rightarrow A \vdash \theta && \text{choice of } \theta.
\end{aligned}
$$

This contradiction completes the proof. ■

The short version of the Theorem 5.3.5 is: Any consistent theory extending N is undecidable, where "undecidable" means not recursive.

Let us apply Theorem 5.3.5 to a particular theory, the theory of the natural numbers, $Th(\mathfrak{N})$.

Theorem 5.3.6 (Tarski's Theorem). *The set of Gödel numbers of formulas true in $\mathfrak{N}$ is not definable in $\mathfrak{N}$.*

Proof: Recall that for any set $A \subseteq \mathbb{N}$, to say ϕ defines A in $\mathfrak{N}$ means that

$$
\begin{aligned}
&\text{If } a \in A, \text{ then } \mathfrak{N} \models \phi(\overline{a}). \\
&\text{If } a \notin A, \text{ then } \mathfrak{N} \models \neg\phi(\overline{a}).
\end{aligned}
$$

Since $Th(\mathfrak{N}) \vdash \alpha$ if and only if $\mathfrak{N} \models \alpha$, we can rewrite this statement as

$$
\begin{aligned}
&\text{If } a \in A, \text{ then } Th(\mathfrak{N}) \vdash \phi(\overline{a}). \\
&\text{If } a \notin A, \text{ then } Th(\mathfrak{N}) \vdash \neg\phi(\overline{a}).
\end{aligned}
$$

So ϕ defines a set in $\mathfrak{N}$ if and only if ϕ represents the set in $Th(\mathfrak{N})$. The set in question is

$$
\begin{aligned}
&\text{TRUEIN}\mathfrak{N} = \\
&\qquad \{a \mid a \text{ is the Gödel number of a formula that is true in } \mathfrak{N}\}.
\end{aligned}
$$

Notice that this is precisely the set

$$\text{THM}_{Th(\mathfrak{N})} =$$
$$\{a \mid a \text{ is the Gödel number of a formula provable from } Th(\mathfrak{N})\},$$

so $\text{TRUEIN}\mathfrak{N}$ is definable if and only if $\text{THM}_{Th(\mathfrak{N})}$ is representable in $Th(\mathfrak{N})$. But $Th(\mathfrak{N})$ is a consistent set of axioms extending N, so by Theorem 5.3.5, $\text{THM}_{Th(\mathfrak{N})}$ is not representable in $Th(\mathfrak{N})$. So $\text{TRUEIN}\mathfrak{N}$ is not definable. ■

5.3.1 Exercises

1. Prove the theorem that is implicit in Definition 5.3.1: If A is a set of sentences, then $\{\sigma \mid \sigma \text{ is a sentence and } A \vdash \sigma\}$ is a theory.

2. Suppose that $\mathfrak{A}$ is an $\mathcal{L}$-structure, and consider $Th(\mathfrak{A})$, as defined in Definition 3.3.4. Prove that $Th(\mathfrak{A})$ is a theory in the sense of Definition 5.3.1.

3. The First Incompleteness Theorem gives us a Π-sentence θ such that $\mathfrak{N} \models \theta$ and $A \nvdash \theta$. Can we find a Σ-sentence with the same characteristics? Please justify your answer.

4. Show that if $X, Y \subset \mathbb{N}$ are recursive sets, then so are $X \cup Y$ and $\mathbb{N} - X$. This fact is used in the comment in the proof of Gödel's First Incompleteness Theorem that the development in Chapter 4 would work for any recursive set of axioms. In Chapter 4 we used heavily the fact that we could write Δ-definitions for everything, but there are recursive sets that do not have Δ-definitions.

5.4 Extensions and Refinements of Incompleteness

If you look carefully at the First Incompleteness Theorem, it does not quite say that the collection of axioms A is incomplete. All that is claimed is that there is a sentence θ such that θ is true-in-$\mathfrak{N}$ and θ is not provable from A. But, perhaps, $\neg\theta$ *is* provable from A. Our first result in this section brings the focus onto incompleteness.

Proposition 5.4.1. *Suppose that A is a consistent, recursive set of axioms that proves all of the axioms of N. If all of the axioms of A are true in $\mathfrak{N}$, then there is a sentence θ such that $A \nvdash \theta$ and $A \nvdash \neg\theta$.*

Proof: As in the proof of the First Incompleteness Theorem, let θ be such that

$$N \vdash \left[\theta \leftrightarrow \neg Thm_A\left(\overline{\ulcorner\theta\urcorner}\right)\right].$$

We know that $\mathfrak{N} \models \theta$ and $A \nvdash \theta$. Suppose that A proves $\neg\theta$. Then, as all of the axioms of A are true in the structure $\mathfrak{N}$, we know that $\mathfrak{N} \models \neg\theta$, which contradicts the fact that $\mathfrak{N} \models \theta$. Thus $A \nvdash \neg\theta$, and A is incomplete. ■

We can also eliminate the hypothesis that the axioms of A be true in $\mathfrak{N}$. To prove that A is incomplete, all that will be required of our axioms will be that they form a consistent extension of N. We will prove this result in two steps, starting by strengthening the hypothesis of consistency to ω-consistency. Then, in Rosser's Theorem, we will show how a slightly trickier use of the Self-Reference Lemma can show that any consistent, recursive extension of N must be incomplete.

Definition 5.4.2. A theory T in $\mathcal{L}_{NT}$ is said to be **ω-inconsistent** if there is a formula $\phi(x)$ such that $T \vdash \exists x \phi(x)$, but for each natural number n, $T \vdash \neg\phi(\overline{n})$. Otherwise, T is called **ω-consistent**.

Proposition 5.4.3. *If T is ω-consistent, then T is consistent.*

Proof: Exercise 2. ■

The converse of this proposition is false, as you are asked to show in Exercise 3. Also notice that if T is a theory such that $\mathfrak{N} \models T$, then T is necessarily ω-consistent. So the chain of implications looks like this:

$$\text{true in } \mathfrak{N} \Rightarrow \omega\text{-consistent} \Rightarrow \text{consistent}.$$

We already know that if A is a recursive, true-in-$\mathfrak{N}$ extension of N, then A is incomplete. In Proposition 5.4.4 we will use the same sentence θ as in the First Incompleteness Theorem to show that an ω-consistent recursive extension of N is incomplete, then in Theorem 5.4.5 we will use a slightly trickier sentence ρ to show that mere consistency suffices: Any consistent, recursive extension of N must be incomplete.

Proposition 5.4.4. *If A is an ω-consistent recursive set of axioms extending N, then A is incomplete.*

Proof: As usual, let θ be such that $N \vdash \left[\theta \leftrightarrow \neg Thm_A\left(\overline{\ulcorner \theta \urcorner}\right)\right]$. We already know that $\mathfrak{N} \models \theta$ and $A \nvdash \theta$. We will show that $A \nvdash \neg\theta$, and thus A is incomplete.

Assume that $A \vdash \neg\theta$; then we know by our choice of θ that $A \vdash Thm_A(\overline{\ulcorner \theta \urcorner})$. In other words,

$$A \vdash (\exists x) Deduction_A(x, \overline{\ulcorner \theta \urcorner}). \tag{5.3}$$

Since we also know that $A \nvdash \theta$, we know, for each natural number n, that n is not the code for a deduction of θ. So

$$\text{For each } n \in \mathbb{N},\ (n, \ulcorner \theta \urcorner) \notin \text{DEDUCTION}_A,$$

and thus, as the formula $Deduction_A$ represents the set DEDUCTION_A,

$$\text{For each } n \in \mathbb{N},\ N \vdash \neg Deduction_A(\overline{n}, \overline{\ulcorner \theta \urcorner}).$$

Since A is an extension of N, we have

$$\text{For each } n \in \mathbb{N},\ A \vdash \neg Deduction_A(\overline{n}, \overline{\ulcorner \theta \urcorner}),$$

which, when combined with (5.3), implies that A is ω-inconsistent, contrary to hypothesis. So our assumption must be wrong, and $A \nvdash \neg\theta$, as needed. ■

We have gotten a lot of mileage out of our sentence θ. Rosser's Theorem uses a different sentence to get a stronger result.

Theorem 5.4.5 (Rosser's Theorem). *If A is a set of $\mathcal{L}_{NT}$-axioms that is recursive, consistent, and extends N, then A is incomplete.*

Proof: We use the Self-Reference Lemma to construct a sentence ρ such that

$$N \vdash \Big[\rho \leftrightarrow$$
$$(\forall x)\left[Deduction_A(x, \overline{\ulcorner \rho \urcorner}) \rightarrow (\exists y < x)(Deduction_A(y, \overline{2}^{\overline{1}}\overline{3}^{\overline{\ulcorner \rho \urcorner}}))\right]\Big].$$

So ρ says, "If there is a proof of ρ, then there is a proof of $\neg\rho$ with a smaller code."

First, we claim that $A \nvdash \rho$: Assume, on the contrary, that $A \vdash \rho$. Let a be a number that codes up a deduction of ρ. Since the formula $Deduction_A$ represents the set DEDUCTION$_A$, we know that

$$A \vdash Deduction_A(\overline{a}, \overline{\ulcorner \rho \urcorner}).$$

Also, by choice of ρ, we know that

$$A \vdash (\forall x)\left[Deduction_A(x, \overline{\ulcorner \rho \urcorner}) \rightarrow (\exists y < x)(Deduction_A(y, \overline{2}^{\overline{1}}\overline{3}^{\overline{\ulcorner \rho \urcorner}}))\right],$$

which means that

$$A \vdash \left[Deduction_A(\overline{a}, \overline{\ulcorner \rho \urcorner}) \rightarrow (\exists y < \overline{a})(Deduction_A(y, \overline{2}^{\overline{1}}\overline{3}^{\overline{\ulcorner \rho \urcorner}}))\right].$$

But we know that $A \vdash Deduction_A(\overline{a}, \overline{\ulcorner \rho \urcorner})$, so we are led to conclude that

$$A \vdash (\exists y < \overline{a})(Deduction_A(y, \overline{2}^{\overline{1}}\overline{3}^{\overline{\ulcorner \rho \urcorner}})). \tag{5.4}$$

On the other hand, we have assumed that $A \vdash \rho$ and A is consistent. Therefore, we know, for each $n \in \mathbb{N}$, that

$$A \vdash \neg Deduction_A(\overline{n}, \overline{2}^{\overline{1}}\overline{3}^{\overline{\ulcorner \rho \urcorner}}).$$

But then, by Rosser's Lemma (Lemma 4.3.7), we know that

$$A \vdash \neg(\exists y < \overline{a})(Deduction_A(y, \overline{2}^{\overline{1}}\overline{3}^{\overline{\ulcorner \rho \urcorner}})),$$

which, when combined with (5.4), shows that A is inconsistent, a contradiction. So we conclude that $A \nvdash \rho$, as claimed.

We also claim that $A \nvdash \neg\rho$. For assume that b is a code for a deduction of $\neg\rho$. Then $A \vdash Deduction_A(\overline{b}, \overline{\ulcorner \neg\rho \urcorner})$. In other words, $A \vdash Deduction_A(\overline{b}, \overline{2}^{\overline{1}}\overline{3}^{\overline{\ulcorner \rho \urcorner}})$.

Now, since $A \vdash \neg\rho$, from the choice of ρ we know that

$$A \vdash (\exists x)\left[Deduction_A(x, \overline{\ulcorner \rho \urcorner}) \land \neg(\exists y)[y < x \land Deduction_A(y, \overline{2}^{\overline{1}}\overline{3}^{\overline{\ulcorner \rho \urcorner}})]\right].$$

So if we substitute $\overline{b}$ for y, we see that

$$A \vdash (\exists x)\left[Deduction_A(x, \overline{\ulcorner \rho \urcorner}) \land \neg[\overline{b} < x \land \underbrace{Deduction_A(\overline{b}, \overline{2}^{\overline{1}}\overline{3}^{\overline{\ulcorner \rho \urcorner}})}_{(*)}]\right].$$

But as the formula (*) is provable from A, this means that

$$A \vdash (\exists x)\left[Deduction_A(x, \overline{\ulcorner \rho \urcorner}) \wedge \neg(\overline{b} < x)\right],$$

which is equivalent to

$$A \vdash (\exists x)\left[x \leq \overline{b} \wedge Deduction_A(x, \overline{\ulcorner \rho \urcorner})\right].$$

Now, since we have assumed that $A \vdash \neg\rho$ and A is consistent, we know that for each $n \in \mathbb{N}$, $A \vdash \neg Deduction_A(\overline{n}, \overline{\ulcorner \rho \urcorner})$, and so by Rosser's Lemma again,

$$A \vdash \neg(\exists x)\left[x \leq \overline{b} \wedge Deduction_A(x, \overline{\ulcorner \rho \urcorner})\right],$$

which shows that A is inconsistent, contrary to assumption. Therefore, $A \nvdash \neg\rho$.

Since $A \nvdash \rho$ and $A \nvdash \neg\rho$, we know that A is incomplete, as needed. ∎

5.4.1 Exercises

1. Suppose that A and B are theories in some language and $A \subseteq B$. Suppose that B is consistent. Show that A is consistent. What happens if B is ω-consistent?

2. Prove Proposition 5.4.3. [*Suggestion:* Try the contrapositive.]

3. Find an example of a theory that is consistent but not ω-consistent. [*Suggestion:* If you can construct the correct sort of model, $\mathfrak{A}$, then $Th(\mathfrak{A})$ will be consistent and ω-inconsistent.]

4. Suppose θ is such that

$$N \vdash \left[\theta \leftrightarrow Thm_A\left(\overline{\ulcorner \neg\theta \urcorner}\right)\right].$$

So θ asserts its own refutability. Is θ true? Provable? Refutable?

5.5 Another Proof of Incompleteness

As we mentioned in the introduction to this chapter, the sentence θ of the First Incompleteness Theorem can be seen as a formalization of the liar paradox, where a speaker assets that what the speaker says is false. In this section we will outline a proof, due to George Boolos [Boolos 94] of the First Incompleteness Theorem that is based upon Berry's paradox.

G. G. Berry, a librarian at Oxford University at the beginning of the twentieth century, is credited by Bertrand Russell with the observation that *the least integer not nameable in fewer than nineteen syllables* is nameable in eighteen syllables. We will formalize a version of Berry's phrase to come up with another sentence that is true in $\mathfrak{N}$ but not provable.

For our argument to work we need to make a minor change in our language. It will be important that our language have only finitely many symbols, and to make $\mathcal{L}_{NT}$ finite, we have to rework the way that we denote variables. So, for this section, rather than having *Vars* be the infinite set of variables $v_1, v_2, \ldots, v_n, \ldots$, and thinking of each v_i as its own symbol, we will think of them as a sequence of symbols. So the string v_{17} is no longer a single symbol but is, rather, three symbols. The set *Vars* then is defined to be the collection of finite strings of symbols that are of the form v_s, where s is a string of numerals. Thus the symbols of $\mathcal{L}_{NT}$ are

$$\{(,), \vee, \neg, \forall, =, v, {}_0, {}_1, {}_2, \ldots, {}_9, 0, S, +, \cdot, E, <\},$$

giving us precisely 23 symbols in the language, and $\mathcal{L}_{NT}$ is finite.

We restate the First Incompleteness Theorem:

Theorem 5.5.1. *Suppose that Q is a consistent and recursive set of $\mathcal{L}_{NT}$-formulas. Then there is a sentence β such that $\mathfrak{N} \models \beta$ and $Q \nvdash \beta$.*

Outline of Proof: We can assume that Q proves all of the axioms of N, since if not, one of the axioms of N would do for β.

Suppose that $\phi(x)$ is a formula of $\mathcal{L}_{NT}$. We say that $\phi(x)$ **names the natural number n with respect to the axioms Q** if and only if

$$Q \vdash \big((\forall x)(\phi(x) \leftrightarrow x = \overline{n})\big).$$

Notice that no formula can name more than one number.

Here is where we use the fact that our language is finite. Fix a number i. Since there are only 23 symbols in our language, there are no more that 23^i formulas of length i, where the length of a formula is the number of symbols that it contains. So no more than 23^i numbers can be named by formulas of length i.

So for each number m, there are only finitely many numbers that can be named by formulas of length less than or equal to m. Thus, for each m, there are some numbers that cannot be named by formulas of length less than or equal to m, so there is a *least* number not named by any formula containing no more than m symbols.

Now there is a formula $\eta(v_1, l)$ with two free variables that says that v_1 is a number named by a formula of length l. You are asked in Exercise 5 to find η. Then we can let $\delta(v_1, v_2)$ be $(\exists l < v_2)\eta(v_1, l)$. Thus $\delta(v_1, v_2)$ says that the number v_1 is nameable in fewer than v_2 symbols.

Two more formulas get us home. Define $\gamma(v_1, v_2)$ by

$$\gamma(v_1, v_2) \text{ is } \left[\neg\delta(v_1, v_2)\right] \wedge \left[(\forall x < v_1)\delta(x, v_2)\right].$$

So $\gamma(v_1, v_2)$ says that v_1 is the least number not nameable in fewer than v_2 symbols. Let k be the number of symbols in $\gamma(v_1, v_2)$. Notice that $k > 4$. (How's that for a bit of an understatement?)

We can now define

$$\alpha(v_1) \text{ is } (\exists v_2)(v_2 = \overline{10} \cdot \overline{k} \wedge \gamma(v_1, v_2)).$$

So $\alpha(v_1)$ claims that v_1 is the least number not nameable in fewer than $10k$ symbols, where k is the number of symbols in $\gamma(v_1, v_2)$. If we write out $\alpha(v_1)$ formally, we see that

$$\alpha(v_1) \text{ is } (\neg(\forall v_2)(\neg(= v_2 \cdot \overline{10}\overline{k} \wedge \gamma(v_1, v_2)))).$$

Let us count the symbols in $\alpha(v_1)$. There are 11 symbols in $\overline{10}$, $k+1$ in $\overline{k}$, and k in $\gamma(v_1, v_2)$. By using both my fingers and my toes, I get a total of $11 + (k+1) + k + 18 = 2k + 30$ symbols in $\alpha(v_1)$. A bit of algebra tells us that since $k > 4$, $10k > 2k + 30$, so $\alpha(v_1)$ contains fewer than $10k$ symbols.

Suppose that b is the smallest number not named by any formula with fewer than $10k$ symbols. Since $\alpha(v_1)$ has fewer than $10k$ symbols, certainly b is not named by the formula $\alpha(v_1)$. By looking back at our definition of what it means for a formula to name a number, we see that

$$Q \nvdash \left((\forall v_1)(\alpha(v_1) \leftrightarrow v_1 = \overline{b})\right).$$

But this is where we want to be. Let β be the sentence

$$\left((\forall v_1)(\alpha(v_1) \leftrightarrow v_1 = \bar{b})\right).$$

We just saw that Q does not prove the sentence β. But β is a true statement about the natural numbers, for the number b *is* the least number that is not nameable in fewer than $10k$ symbols, and that is precisely the meaning of the sentence β. So $\mathfrak{N} \models \beta$ and $Q \nvdash \beta$, as needed. ■

One big difference between this proof and our first proof of the First Incompleteness Theorem is that this proof does not use the Self-Reference Lemma, as we don't have to substitute the Gödel number of a formula into itself, but rather, we substitute the numeral of a number that makes the formula true. But both proofs do rely on Gödel numbering and coding deductions, so the mechanism of Chapter 4 comes into play for both.

5.5.1 Exercises

1. Give an argument, perhaps based on Church's Thesis or perhaps using a variant of Theorem 4.10.2, to show that there is a Δ-formula *LengthOfFormula*(f, l) that is true if and only if f is the Gödel number of a formula consisting of exactly l symbols. [*Suggestion:* You may need to start by showing the existence of a formula *LengthOfTerm* with two free variables.]

2. Prove that no formula can name two different numbers. [*Suggestion:* Think about what it would mean if $\phi(x)$ named both 17 and 42.]

3. Find an upper bound for the number of numbers that can be named by formulas $\phi(x)$ that contain no more than m symbols.

4. Suppose that $\phi(x)$ names n with respect to the set of axioms N. Show that $\phi(x)$ represents $\{n\}$.

5. Find a formula $\eta(n, l)$ that says that n is named by a formula of length l. Is your formula equivalent to a Σ-formula?

6. The statement of Theorem 5.5.1 makes two assumptions about the collection of axioms Q. Where are they used in the proof?

5.6 Peano Arithmetic and the Second Incompleteness Theorem

It is our goal in this section to show that a set of axioms cannot prove its own consistency. Now this statement needs to be sharpened, for of course *some* sets of axioms can prove their own consistency. For example, the axiom set A might contain the statement "A is consistent." But that will lead to problems, as we will show.

The first order of business will be to introduce a new collection of axioms, called PA, or the axioms of Peano Arithmetic. This extension of N will be recursive and will be true in $\mathfrak{N}$, so all of the results of this chapter will apply to PA. We will then state, without proof, three properties that are true of PA, properties that are needed for the proof of the Second Incompleteness Theorem.

The Second Incompleteness Theorem is, in some sense, nothing more than finding another true and unprovable statement, but the statement that we will find is much more natural and has a longer history than the sentence θ of Gödel I. As we mentioned on page 1, at the beginning of the twentieth century, the German mathematician David Hilbert proposed that the mathematical community set itself the goal of proving that mathematics is consistent. In the Second Incompleteness Theorem, we will see that no extension of Peano Arithmetic can prove itself to be consistent, and thus certainly any plan for a self-contained proof of consistency must be doomed to failure. This was the blow that Gödel delivered to Hilbert's consistency program. Understanding the ideas behind this second proof is our current goal. We will not fill in all of the details of the construction. The interested reader is directed to Craig Smoryński's article in [Barwise 77], on which our presentation is based.

We begin by establishing our new set of nonlogical axioms, the axioms of Peano Arithmetic.

Definition 5.6.1. The axioms of **Peano Arithmetic**, or PA, are the eleven axioms of N together with the axiom schema

$$\left[\phi(0) \land (\forall x)\left[\phi(x) \rightarrow \phi(Sx)\right]\right] \rightarrow (\forall x)\phi(x)$$

for each $\mathcal{L}_{NT}$-formula ϕ with one free variable.

So Peano Arithmetic is nothing more than the familiar set of axioms N, together with an induction schema for $\mathcal{L}_{NT}$-definable sets.

Although we will not go through the details, it is easy to see that the set AXIOMOFPA is recursive, so *PA* is a recursively axiomatized extension of N.

What makes *PA* useful to us is that *PA* is strong enough to prove certain facts about derivations in *PA*. In particular, *PA* is strong enough so that the following derivability conditions hold for all formulas ϕ:

$$\text{If } PA \vdash \phi, \text{ then } PA \vdash Thm_{PA}(\overline{\ulcorner \phi \urcorner}). \tag{D1}$$

$$PA \vdash \left[Thm_{PA}(\overline{\ulcorner \phi \urcorner}) \rightarrow Thm_{PA}\left(\overline{\ulcorner Thm_{PA}(\overline{\ulcorner \phi \urcorner}) \urcorner}\right) \right]. \tag{D2}$$

$$PA \vdash \left[\left[Thm_{PA}(\overline{\ulcorner \phi \urcorner}) \wedge Thm_{PA}(\overline{\ulcorner \phi \rightarrow \psi \urcorner}) \right] \rightarrow Thm_{PA}(\overline{\ulcorner \psi \urcorner}) \right]. \tag{D3}$$

Granting these conditions, we can move on to prove the Second Incompleteness Theorem. Recall that we agreed to use the symbol $\perp$ for the contradictory sentence $\left[(\forall x)x = x\right] \wedge \neg\left[(\forall x)x = x\right]$.

Definition 5.6.2. The sentence Con_{PA} is the sentence $\neg Thm_{PA}(\overline{\ulcorner \perp \urcorner})$.

Notice that $\mathfrak{N} \models Con_{PA}$ if and only if *PA* is a consistent set of axioms. For if *PA* is not consistent, then *PA* can prove anything, including $\perp$. If *PA* is consistent, since we know that there is a proof in *PA* of $\neg \perp$, there must not be a proof of $\perp$.

Theorem 5.6.3 (Gödel's Second Incompleteness Theorem). *If Peano Arithmetic is consistent, then* $PA \nvdash Con_{PA}$.

Proof: Let θ be, as usual, the statement generated by the Self-Reference Lemma, but this time we apply the lemma to the formula $\neg Thm_{PA}(v_1)$. So θ is such that

$$PA \vdash \left[\theta \leftrightarrow \neg Thm_{PA}\left(\overline{\ulcorner \theta \urcorner}\right) \right]. \tag{5.5}$$

We know that $PA \nvdash \theta$, as *PA* is a recursive consistent extension of N, all of whose axioms are true in $\mathfrak{N}$ (Proposition 5.4.1). We will show that

$$PA \vdash (\theta \leftrightarrow Con_{PA}).$$

If it was the case that $PA \vdash Con_{PA}$, then we would have $PA \vdash \theta$, a contradiction. Thus $PA \nvdash Con_{PA}$. [For this proof, all we really need is that $PA \vdash \left(Con_{PA} \rightarrow \theta\right)$, but the other direction is used in the Exercises.]

So all that is left is to show that $PA \vdash \left(\theta \leftrightarrow Con_{PA}\right)$.

For the forward direction, since $\perp$ is the denial of a tautology, we know that

$$PA \vdash (\perp \rightarrow \theta),$$

so by the first derivability condition (D1) we know that

$$PA \vdash Thm_{PA}\left(\overline{\ulcorner \perp \rightarrow \theta \urcorner}\right).$$

This implies, via (D3), that

$$PA \vdash Thm_{PA}(\overline{\ulcorner \perp \urcorner}) \rightarrow Thm_{PA}(\overline{\ulcorner \theta \urcorner}),$$

which is equivalent to

$$PA \vdash \neg Thm_{PA}(\overline{\ulcorner \theta \urcorner}) \rightarrow \neg Thm_{PA}(\overline{\ulcorner \perp \urcorner}). \tag{5.6}$$

Now, if we combine (5.5) and (5.6), we see that

$$PA \vdash \theta \rightarrow \neg Thm_{PA}(\overline{\ulcorner \perp \urcorner}),$$

which is equivalent to

$$PA \vdash \theta \rightarrow Con_{PA},$$

which is half of what we need to prove.

For the converse, notice that from derivability condition (D2),

$$PA \vdash Thm_{PA}(\overline{\ulcorner \theta \urcorner}) \rightarrow Thm_{PA}\left(\overline{\ulcorner Thm_{PA}(\overline{\ulcorner \theta \urcorner}) \urcorner}\right). \tag{5.7}$$

Since we also know that $PA \vdash \theta \leftrightarrow \neg Thm_{PA}(\overline{\ulcorner \theta \urcorner})$, the sentence

$$Thm_{PA}\left(\overline{\ulcorner Thm_{PA}(\overline{\ulcorner \theta \urcorner}) \rightarrow \neg\theta \urcorner}\right)$$

is a true Σ-sentence. Since N suffices to prove true Σ-sentences, certainly

$$PA \vdash Thm_{PA}\left(\overline{\ulcorner Thm_{PA}(\overline{\ulcorner \theta \urcorner}) \rightarrow \neg\theta \urcorner}\right). \tag{5.8}$$

Now, if we take (5.8) and derivability condition (D3), we have

$$PA \vdash Thm_{PA}\left(\overline{\ulcorner Thm_{PA}(\overline{\ulcorner \theta \urcorner}) \urcorner}\right) \rightarrow Thm_{PA}(\overline{\ulcorner \neg\theta \urcorner}). \tag{5.9}$$

If we combine (5.7) and (5.9), we see that

$$PA \vdash Thm_{PA}(\overline{\ulcorner \theta \urcorner}) \rightarrow Thm_{PA}(\overline{\ulcorner \neg\theta \urcorner}). \tag{5.10}$$

Now, since we know that the sentence $\theta \rightarrow \left[(\neg\theta) \rightarrow \perp\right]$ is a tautology, the statement

$$Thm_{PA}\left(\overline{\ulcorner \theta \rightarrow \left[(\neg\theta) \rightarrow \perp \urcorner\right.}\right)$$

is a true Σ-sentence, so

$$PA \vdash Thm_{PA}\left(\overline{\ulcorner \theta \rightarrow \left[(\neg\theta) \rightarrow \perp \urcorner\right.}\right). \tag{5.11}$$

Once again, using the derivability condition (D3) twice on (5.11), we find that

$$PA \vdash Thm_{PA}(\overline{\ulcorner \theta \urcorner}) \rightarrow \left[Thm_{PA}(\overline{\ulcorner \neg\theta \urcorner}) \rightarrow Thm_{PA}(\overline{\ulcorner \perp \urcorner})\right],$$

and if we combine that with (5.10), we see that

$$PA \vdash Thm_{PA}(\overline{\ulcorner \theta \urcorner}) \rightarrow Thm_{PA}(\overline{\ulcorner \perp \urcorner}),$$

which is equivalent to

$$PA \vdash \neg Thm_{PA}(\overline{\ulcorner \perp \urcorner}) \rightarrow \neg Thm_{PA}(\overline{\ulcorner \theta \urcorner}),$$

which when we translate the antecedent and use the definition of θ on the consequent, combines with (D3) to give us

$$PA \vdash Con_{PA} \rightarrow \theta,$$

which is what we needed to prove. ■

The proof of the Second Incompleteness Theorem is technical, but the result is fabulous: If Peano Arithmetic is consistent, it cannot prove its own consistency. You will not be surprised to find out that the same result holds for any consistent set of axioms extending Peano Arithmetic that can be represented by a Σ-formula.

> *Chaff:* Time for a bit of technical stuff that is pretty neat. To be precise, the way in which we code up the axioms of Peano Arithmetic is important in the statement of the Second Incompleteness Theorem. The set AXIOMOFPA is recursive, and thus there is a formula $\phi(x)$ that represents that set. In fact, the formula $\phi(x)$ can be taken to be a Δ-formula, and if you use *that* ϕ, then Gödel's result is as we have stated it. But there are other formulas that you could use to represent the set of axioms of Peano Arithmetic, and Solomon Feferman proved in 1960 that it is possible to express consistency using one of these other formulas in such a way that $PA \vdash Con_{PA}$ [Feferman 60]. The moral is: You have to be very precise when dealing with consistency statements.

The example of Con_{PA} as a true and yet unprovable statement stood for 45 years as the most natural sentence of that type. In 1977, however, Jeff Paris and Leo Harrington discovered a generalization of Ramsey's Theorem in combinatorics that is not provable in Peano Arithmetic. Since that time a small cottage industry has developed that produces relatively natural statements that are not provable in one theory or another.

We end this section with a couple of interesting corollaries of the Second Incompleteness Theorem. The first corollary is pretty strange. Suppose for a second that Peano Arithmetic is consistent (you believe that it is, right?). Then we can, if we like, assume that PA is *inconsistent*, and we will still have a consistent set of axioms!

Corollary 5.6.4. *If PA is consistent, the set of axioms*

$$PA \cup \{\neg Con_{PA}\}$$

is consistent.

Proof: This is immediate. We know that $PA \nvdash Con_{PA}$. So by Exercise 4 in Section 2.7.1, $PA \cup \{\neg Con_{PA}\}$ is consistent. ■

To appreciate the next corollary, remember our development of the First Incompleteness Theorem, in the setting of Peano Arithmetic. The sentence θ that we construct from the Self-Reference Lemma "says" that it (θ) is not provable from PA. Then the Incompleteness Theorem says that θ is right: θ is unprovable from PA, so θ is correct in what it asserts.

Now, suppose that we use the Self-Reference Lemma to construct a different sentence, α, such that α claims that it *is* provable from PA. Is α true? False? Löb's Theorem provides the answer.

Corollary 5.6.5 (Löb's Theorem). *Suppose α is such that $PA \vdash Thm_{PA}(\overline{\ulcorner \alpha \urcorner}) \rightarrow \alpha$. Then $PA \vdash \alpha$.*

Proof: This proof hinges on the fact that PA is sufficiently strong to prove the equivalence

$$\neg Thm_{PA}(\overline{\ulcorner \alpha \urcorner}) \leftrightarrow Con_{PA \cup \{\neg \alpha\}}.$$

Granting this, since by assumption $PA \vdash \left(\neg \alpha \rightarrow \neg Thm_{PA}(\overline{\ulcorner \alpha \urcorner})\right)$, we know that

$$PA \cup \{\neg \alpha\} \vdash \neg Thm_{PA}(\overline{\ulcorner \alpha \urcorner}),$$

so by our equivalence,

$$PA \cup \{\neg \alpha\} \vdash Con_{PA \cup \{\neg \alpha\}}.$$

But then by the Second Incompleteness Theorem, $PA \cup \{\neg \alpha\}$ must be inconsistent, and therefore $PA \vdash \alpha$, as needed. ■

Löb's Theorem tells us that the sentence that states, "I am provable in Peano Arithmetic" is true (and therefore provable)!

5.6.1 Exercises

1. Explain how the induction schema of Peano Arithmetic as given in Definition 5.6.1 differs from the full principle of mathematical induction:

$$(\forall A \subseteq \mathbb{N}) \Big[\big(0 \in A \land (\forall x)(x \in A \rightarrow Sx \in A)\big) \rightarrow A = \mathbb{N} \Big].$$

[*Suggestion:* You might look at the comment on page 118.]

2. Suppose θ and η are two sentences that assert their own unprovability (from PA). So

$$PA \vdash \left[\theta \leftrightarrow \neg Thm_{PA}\left(\overline{\ulcorner \theta \urcorner}\right)\right]$$

and

$$PA \vdash \left[\eta \leftrightarrow \neg Thm_{PA}\left(\overline{\ulcorner \eta \urcorner}\right)\right].$$

Prove that $PA \vdash \theta \leftrightarrow \eta$.

3. Let ρ be the Rosser sentence (in the setting of PA). So

$$PA \vdash \Big[\rho \leftrightarrow (\forall x)\big[Deduction_{PA}(x, \overline{\ulcorner\rho\urcorner}) \rightarrow (\exists y < x)(Deduction_{PA}(y, \overline{2}^{\overline{1}}\overline{3}^{\overline{\ulcorner\rho\urcorner}}))\big]\Big].$$

Show that

$$PA \vdash \Big[Con_{PA} \rightarrow \big[(\neg Thm_{PA}(\overline{\ulcorner\rho\urcorner})) \wedge (\neg Thm_{PA}(\overline{\ulcorner\neg\rho\urcorner}))\big]\Big].$$

Use this and Gödel's Second Incompleteness Theorem to show

$$PA \vdash \big[Con_{PA} \rightarrow \rho\big]$$

and

$$PA \nvdash \big[\rho \rightarrow Con_{PA}\big].$$

So since Gödel's θ is such that $PA \vdash [\theta \leftrightarrow Con_{PA}]$, you have shown that θ and ρ are not provably equivalent in PA.

5.7 George Boolos on the Second Incompleteness Theorem

George Boolos published a wonderful article on the Second Incompleteness Theorem. Here is the first page of his

Gödel's Second Incompleteness Theorem Explained in Words of One Syllable [1]

First of all, when I say "proved", what I will mean is "proved with the aid of the whole of math". Now then: two plus two is four, as you well know. And, of course, *it can be proved* that two plus two is four (proved, that is, with the aid of the whole of math, as I said, though

[1] *Mind*, Vol. 103, January 1994, pp. 1–3. Used with the permission of Oxford University Press.

in the case of two plus two, of course we do not need the *whole* of math to prove that it is four). And, as may not be quite so clear, it can be proved that it can be proved that two plus two is four, as well. And it can be proved that it can be proved that it can be proved that two plus two is four. And so on. In fact, if a claim can be proved, then it can be proved that the claim can be proved. And *that* too can be proved.

Now: two plus two is not five. And it can be proved that two plus two is not five. And it can be proved that it can be proved that two plus two is not five, and so on.

Thus: it can be proved that two plus two is not five. Can it be proved as well that two plus two *is* five? It would be a real blow to math, to say the least, if it could. If it could be proved that two plus two is five, then it could be proved that five is not five, and then there would be *no* claim that could *not* be proved, and math would be a lot of bunk.

So, we now want to ask, can it be *proved* that it can't be proved that two plus two is five? Here's the shock: no, it can't. Or, to hedge a bit: *if* it can be proved that it can't be proved that two plus two is five, *then* it can be proved as well that two plus two is five, and math is a lot of bunk. In fact, if math is not a lot of bunk, then no claim of the form "claim X can't be proved" can be proved.

So, if math is not a lot of bunk, then, though it can't be proved that two plus two is five, it can't be proved *that* it can't be proved that two plus two is five.

By the way, in case you'd like to know: yes, it *can* be proved that if it can be proved that it can't be proved that two plus two is five, then it can be proved that two plus two is five.

5.8 Summing Up, Looking Ahead

This chapter is the capstone of the book. We have stated and proved (with only a small bit of handwaving) the two incompleteness theorems. I am sure that at this point you have a strong understanding of the theorems as well as an appreciation for the delicate arguments that are used in their proofs. These theorems have had a profound impact on the philosophical understanding of the subject of mathematics, and they involve some wonderful mathematics in and of themselves.

When I was in college, I was constantly asked by my mother what I was studying, and it became sort of a game to try to distill the essence of a course into a few phrases that would get the idea across without bogging down into technical details. As I think through this course, I think the summary I would give my mom would be something like this:

> We looked at the language in which mathematics is written and looked at different kinds of mathematical universes, or structures. We thought about axioms and proofs. We worked through the proof of Gödel's Completeness Theorem, which shows that any set of axioms that is not self-contradictory is true in some mathematical universe.
>
> Then we changed our focus and thought about the natural numbers and ordinary arithmetic. We proved that $1+1$ is 2. We proved that we can prove that $1+1$ is 2. But the amazing thing is that even though $1+1$ is not 3, and even though we can prove that $1+1$ is not 3, we *cannot* prove that we cannot prove that $1+1$ *is* 3, unless mathematics is inconsistent. That is the core of Gödel's Second Incompleteness Theorem.

So five years from now, if you have stopped studying mathematics and think back on this course, I hope you will at least remember the core ideas. If, on the other hand, you are actively working in a mathematical field, I hope that you will also remember some of the intricate arguments that we have developed and some of the powerful theorems that we have seen. Logic is a useful and powerful part of mathematics, interesting in its own right and shaping our understanding of the entire field.

Appendix Just Enough Set Theory to Be Dangerous

It is my goal in this appendix to review some basic set-theoretic notions and to state some results that are used in the book. Very little will be proved, but the Exercises will give you a chance to get a feel for the subject. There are several laudable texts and reference works on set theory, I have listed a few in the Bibliography.

Think of a set as a collection of objects. If X is a set, we write $a \in X$ to say that the object a is in the collection X. For our purposes, it will be necessary that any given thing either is an element of a given set or is not an element of that set. If that sounds obvious to you, suppose I asked you whether or not 153,297 was in the "set" of big numbers. Depending on your age and whether or not you are used to handling numbers that large, your answer might be "yes," "no," or "pretty large, but not all that large." So you can see that membership in an alleged set might not be all that cut and dried. But we won't think about sets of that sort, leaving them to the field of fuzzy set theory.

Our main concern will be with infinite sets and the size of those sets. One of the really neat results of set theory is that infinite sets come in different sizes, so there are different sizes of infinity! Here are some of the details:

We say that sets X and Y have the same **cardinality**, and write $|X| = |Y|$, if there is a one-to-one and onto function (also called a *bijection*) $f : X \to Y$. The function f is sometimes called a *one-to-one correspondence.* For example, if X and Y are finite sets, this just means that you can match up the elements of the set without missing anyone. So if $D = \{$Happy, Sleepy, Grumpy, Doc, Dopey, Sneezy, Bashful$\}$ and $W = \{$Sunday, Monday, Tuesday, Wednesday, Thursday, Friday, Saturday$\}$, it is easy to see that $|D| = |W|$ by pairing them up. Notice in doing the pairing that you never actually have to count the number of elements in either set. All you have to do is match them up. This is what allows us to apply the concept of cardinality to infinite sets.

As an easy example, notice that $\mathbb{N} = \{0, 1, 2, 3, \ldots\}$ has the same cardinality as $\textsc{Even} = \{0, 2, 4, 6, \ldots\}$. To prove this, I must exhibit a bijection between the two sets, and the function $f : \mathbb{N} \to \textsc{Even}$ defined by $f(x) = 2x$ works quite nicely. It is easy to check that f is a bijection, so $|\textsc{Even}| = |\mathbb{N}|$. Notice that this says that these two sets have the same size, even though the first is a proper subset of the second. This cannot happen with finite sets, but is rather common with infinite sets.

A set that has the same cardinality as the set of natural numbers is called **countable**. The exercises will give you several opportunities to show that the cardinality of one set is equal to the cardinality of another set.

We will say that the cardinality of set A is less than or equal to the cardinality of set B, and write $|A| \leq |B|$, if there is a one-to-one function $f : A \to B$. If $|A| \leq |B|$ but $|A| \neq |B|$, we say that the cardinality of A is less than the cardinality of B.

One of the basic theorems of set theory, the Schröder–Bernstein Theorem, says that if $|A| \leq |B|$ and $|B| \leq |A|$, then $|A| = |B|$. The proof of this theorem is beyond this little appendix, but a nice version is in [Hrbacek and Jech 84]. Be warned, however. This theorem looks obvious, because of the notation. What it says, however, is that if you have two sets A and B such that there are injections $f_1 : A \to B$ and $f_2 : B \to A$, then there is a bijection $f_3 : A \to B$. But for our purposes, it will suffice to think: "If the size of this set is no bigger than the size of that set, and if the size of that set is no bigger than the size of this set, then the two sets must have the same size!"

With this tool in hand it is not hard to show that $\mathbb{Q}$, the collection of rational numbers, is countable. An injection from $\mathbb{N}$ to $\mathbb{Q}$ is easy: $f(x) = x$ will do. But to find an injection from the rationals to the naturals requires more thought. I'll let you think about it, but trust me when I say that there is such a function. If you don't trust me, check any introductory set theory book or discrete mathematics textbook.

You will have noticed by this point that all we have done is show that several different infinite sets are countable. Georg Cantor proved in 1874 that the set of real numbers is not countable, in other words, he showed that $|\mathbb{N}| < |\mathbb{R}|$. In 1891 he developed his famous diagonal argument to show that given any set X, there is a set of larger cardinality, namely the collection of all subsets of X, which we call the *power set* of X and denote $P(X)$. So states $|X| < |P(X)|$. Thus there can be no largest cardinality. More picturesquely: There is no largest set.

The way to think about a countable set is that you can write an infinite list that contains every element of the set. This means that you can go through the set one element at a time and be sure that you eventually get to every element of the set, and that any element of the set only has finitely many things preceding it on the list. Think about listing the natural numbers. Although it would take a long time to reach the number 1038756397456 on the list $\langle 0, 1, 2, 3, 4, \ldots \rangle$, we would eventually get to that number, and after only a finite amount of time. The fact that the real numbers are uncountable means that if you try to write them out in a list, you have to leave some reals off your list. So if you try to list the reals as $\langle 1, -4/7, e, 10^{10}, \pi/4, \ldots \rangle$, then there has to be a real number, maybe 17, that you left out.

Suppose that X_1, X_2, X_3, $\ldots$ are all countable sets, and consider the set $Y = X_1 \cup X_2 \cup X_3 \cup \cdots$. So Y is a countable union of countable sets. If you want to know the size of Y, it is easy to find, for there is a theorem that states that the countable union of countable sets is countable. (Purists: Calm down. I know about the Axiom of Choice. I know we don't need this theorem in this generality. But this isn't a set theory text, so lighten up a little!) Let's look at an application of this wonderful theorem:

Almost all of the languages in this book are countable, meaning that the set consisting of all of the constant symbols, function symbols, and relation symbols, together with the connectives, parentheses, and variables is countable. Now using the fact that a countable union of countable sets is countable, we can show that S_2, the collection of all strings of two symbols from the language, is countable.

Here is a table that contains every string of exactly two $\mathcal{L}$-symbols, where $\mathcal{L} = \langle s_0, s_1, s_2, \dots \rangle$:

	s_0	s_1	s_2	s_3	s_4	$\cdots$
s_0	s_0s_0	s_0s_1	s_0s_2	s_0s_3	s_0s_4	
s_1	s_1s_0	s_1s_1	s_1s_2	s_1s_3	s_1s_4	$\cdots$
s_2	s_2s_0	s_2s_1	s_2s_2	s_2s_3	s_2s_4	
s_3	s_3s_0	s_3s_1	s_3s_2	s_3s_3	s_3s_4	
			$\vdots$			

Now, this array shows that S_2 can be thought of as a countable union of countable sets: Each row of the table is countable, and there are countably many rows. By the theorem of the last paragraph, the collection of length-two strings is countable.

> *Chaff:* Notice that by reading the table diagonally starting in the upper left-hand corner, we can explicitly construct a listing of all of the length-two strings:
>
> $$\langle s_0s_0, s_0s_1, s_1s_0, s_0s_2, s_1s_1, s_2s_0, s_0s_3, s_1s_2, s_2s_1, s_3s_0, \dots \rangle.$$
>
> This trick is important when you want to be able to program a computer to produce the listing or when you want to show that the collection of length two strings is recursive.

Using the same idea, we can prove by induction that for any natural number n, S_n, the collection of strings of length n is countable. [Put the length $(n-1)$ strings across the top of the table and the symbols of $\mathcal{L}$ down the side.] But then the collection of *all* finite strings is just the (countable) union $S_0 \cup S_1 \cup S_2 \cup \cdots$. So the collection of all finite strings is countable. Exercise 5 asks you to think about generating the listing of strings via a computer.

Exercises

1. Show that the set of numbers that are a nonnegative power of 10 (also known as $A = \{1, 10, 100, 1000, \dots\}$) is countable.

2. Show that the collection of prime numbers is countable.

3. (a) Show that these two intervals on the real line have the same cardinality: $[0, 1]$ and $[0, 10]$.

(b) Show that these two intervals have the same cardinality: $[0, 1]$ and $[1, 3]$.

(c) Show that these two intervals on the real line have the same cardinality: $[a, b]$ and $[c, d]$, where $a \neq b$ and $c \neq d$.

(d) Show that the interval $[0, 1]$ and the interval $(0, 1)$ have the same cardinality. [*Suggestion:* The easy way is to quote a theorem. The interesting way is to find an explicit bijection between the two sets.]

4. Show that the relation "has the same cardinality as" is an equivalence relation.

5. Suppose that we have a recursively enumerable listing of the elements of a countable language $\mathcal{L} = \langle s_0, s_1, s_2, \ldots \rangle$. Design a computer program that lists all of the finite strings of symbols from the language, showing that the collection of finite strings is also recursively enumerable.

Bibliography

[Barwise 77] Jon Barwise, ed. *Handbook of Mathematical Logic*. Amsterdam: North-Holland, 1977.

[Bell and Machover 77] John L. Bell and Moshé Machover. *A Course in Mathematical Logic*. Amsterdam: North-Holland, 1977.

[Boolos 89] George Boolos. A New Proof of the Gödel Incompleteness Theorem. *Notices of the American Mathematical Society,* Vol. 36, No. 4, April 1989, pp. 388–390.

[Boolos 94] ——— Gödel's Second Incompleteness Theorem Explained in Words of One Syllable. *Mind,* Vol. 103, January 1994, pp. 1–3.

[Chang and Keisler 73] C. C. Chang and H. J. Keisler. *Model Theory*. Amsterdam: North-Holland, 1973.

[Crossley et al. 72] J. N. Crossley, C. J. Ash, C. J. Brickhill, J. C. Stillwell, and N. H. Williams. *What Is Mathematical Logic?* London: Oxford University Press, 1972.

[Dawson 97] John W. Dawson, Jr. *Logical Dilemmas: The Life and Work of Kurt Gödel*. Wellesley, Mass.: A. K. Peters, 1997.

[Enderton 72] Herbert B. Enderton. *A Mathematical Introduction to Logic*. Orlando, Fla.: Academic Press, 1972.

[Feferman 60] Solomon Feferman. Arithmetization of Metamathematics in a General Setting. *Fundamenta Mathematicae,* Vol. 49, 1960, pp. 35–92.

[Feferman 98] ——— *In the Light of Logic*. Oxford: Oxford University Press, 1998.

[Gödel–Works] Kurt Gödel. *Collected Works:* Vol. I, *Publications 1929–1936.* Edited by Soloman Feferman, John W. Dawson, Jr., Stephen C. Kleene, Gregory H. Moore, Robert M. Solovay, and Jean Van Heijenoort. New York: Oxford University Press, 1986.

[Goldstern and Judah 95] Martin Goldstern and Haim Judah. *The Incompleteness Phenomenon: A New Course in Mathematical Logic.* Wellesley, Mass.: A. K. Peters, 1995.

[Henle 86] James M. Henle. *An Outline of Set Theory.* New York: Springer-Verlag, 1986.

[Hrbacek and Jech 84] Karel Hrbacek and Thomas Jech. *Introduction to Set Theory.* New York: Marcel Dekker, 1984.

[Keisler 76] H. Jerome Keisler. *Foundations of Infinitesimal Calculus.* Boston: Prindle, Weber & Schmidt, 1976.

[Keisler and Robbin 96] H. Jerome Keisler and Joel Robbin. *Mathematical Logic and Computability.* New York: McGraw-Hill, 1996.

[Malitz 79] Jerome Malitz. *Introduction to Mathematical Logic.* New York: Springer-Verlag, 1979.

[Manin 77] Yu. I. Manin. *A Course in Mathematical Logic.* New York: Springer-Verlag, 1977.

[Mendelson 87] Elliott Mendelson. *Introduction to Mathematical Logic.* Monterey, Calif.: Brooks/Cole, a division of Wadsworth, 1987.

[Roitman 90] Judith Roitman. *Introduction to Modern Set Theory.* New York: Wiley, 1990.

[Russell 67] Bertrand Russell. *The Autobiography of Bertrand Russell,* Vol. I. London: George Allen & Unwin, 1967.

Index

W